The New Physics of Consciousness

Reconciling Science & Spirituality

"Science without religion is lame. Religion without science is blind"

Albert Einstein

The book includes a true David and Goliath story. In a revolutionary approach to science David makes a deadly challenge to a major pillar of quantum theory. Albert Einstein described this pillar as *"...a real witches calculus...most ingenious, and adequately protected by its great complexity against being proved wrong."* If David vindicates Einstein, physics teaching in every university of the world would be impacted. Nobel Prizes may be questioned. Even the scientific method could collapse. Nietzsche declared; *God is dead.* If Einstein is proved right it will not be God declared dead but science as we know it!

Other Books by David Ash:
The Tower of Truth (CAMSPRESS, Cornwall 1977),
Science of The Gods/ The Vortex: Key to Future Science – with Peter Hewitt - (Gateway Books, Bath, 1990),
The Science of Ascension (Cloverleaf Connection, Canada, 1993),
Activation for Ascension (Kima Global Publishing, Cape town RSA, 1995)
The *New Science of the Spirit* (The College of Psychic Studies, London, 1995)
God the Ultimate Paradox (Kima Global Publishing, Cape town RSA, 1998)
Is God Good? (Kima Global Publishing, Cape town RSA, 2005)
The Role of Evil in Human Consciousness (Kima Global Publishers, RSA, Cape Town, 2007)

The New Physics of Consciousness

Reconciling Science & Spirituality

David Ash

Published by Kima Global Publishers
Kima Global House,
50, Clovelly Road
Clovelly 7975
P.O. Box 374,
Rondebosch 7701
South Africa

ISBN: 978-0-9802561-2-3
Cover design: Nadine May

e-mail: info@kimaglobal.co.za
website: www.kimaglobal.co.za
Author's website: www.davidash.info

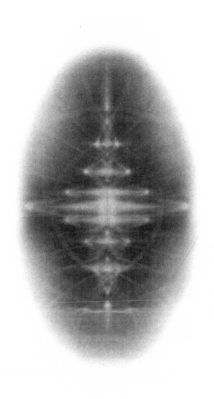

To Amma

Water in its turbulent hilarity,
Creates whirlpool forms,
Which have no independent reality,
In the twinkling of an eye,
They would vanish if the river ran dry.

All of man's material things,
For which he fights and to which he clings,
Formed by light when it spins,
Like whirlpools in busy streams,
Are no more than vacant dreams.

Light in its ceaseless whirling race,
Creates the forms of matter and space,
From existence the Universe would vanish,
Even the void would be banished,
Should light cease to spin.

Acknowledgements

"Lo you creative being, what rich out flowing you're enabling! On behalf of everyone who will be inspired, delighted and enlightened by your book...Thankyou!"

Emily Longhurst

I met Emily at the Crystal Palace, London in 2005 when we both went to receive our first hugs from Amma. A year later I was in the middle of a divorce and surrounded by letters of rejection from publishers when, out of the blue, Emily said she loved me because of my tenacity and determination to continue with my work in physics despite setbacks, rejections and lack of recognition. It only took the love of one woman. On her birthday, October 19, 2006 this book was born. The introduction, first two chapters and a synopsis were in the post to a publisher before my birthday on the 20th. After the rejection arrived it was for love of Emily that I finished *The New Physics of Consciousness*. She drew it out of me with her love so I wrote it just for her. This is a true love story.

Yvonne appeared in November at the Amma programme in Dublin with heaven in her eyes and more words of upliftment. From the generation of my daughters, Emily and Yvonne were calling me to complete my mission and get this work out. Heaven has its way for dreams to be realised.

Hillary took me to Amma who hugs with boundless love and teaches that God is consciousness. Words cannot describe my love for Amma. Anna, Hillary and Liesbeth were supportive of my work and my sons and daughters and Liesbeth's daughters love me come what may. Colin and Miranda Robertson and John and Mary Bingham provided me with inspiration, words of wisdom and safe havens to research and write my books. My brothers and sisters encouraged me with humour. Anna and our daughters drew the hedgehogs, tadpoles, elephants and ants, Mr. Proton, Victor Schauberger and Democritus.

I am ever grateful to my dear friend and kindred spirit Robin Beck for publishing this book and Peter Hewitt, Sir George Trevelyan, Nigel Blair, David Bosworth, Colin Wilson and Leslie Kenton for their belief in my work. Finally I offer heart felt gratitude to my father, Michael Ash for rearing me for this task and my mother, Joy Ash for her unwavering love and encouragement.

Contents

Introduction

Today there is a wide measure of agreement that the stream of knowledge is heading toward a non-mechanical reality; the universe begins to look more like a great thought than a great machine.

Sir James Jeans

If modern science were based on ancient Indian philosophy rather than ancient Greek philosophy we would be living in a different world with different values. When philosophers in ancient Greece decided the atom was a material particle they were relying on the meandering of their minds in the mires of speculation. Yogis in ancient India had the siddhi power to direct their minds into the atom and see what was going on in there. What they spotted was 'the smallest particles of matter are vortices of energy.'

Imagining atoms as material particles the Greeks invented materialism. Seeing the vortex in the atom the Yogis recognised maya; they realised the vortex creates the illusion of material.

Treating subatomic particles as vortices of energy enabled me to explain everything in physics in terms of a single principle and publish as a prediction, in 1995, that the more distant galaxies are from us, the faster they move away from us. This prediction of the accelerating expansion of the Universe was confirmed in 1998 by observations of supernova explosions in distant galaxies.

A theory that explains everything from a single principle is one thing. When it is proved by predicting the greatest astronomical discovery of the closing decade of the twentieth century; that is something else. The implication of a proven physics based on Indian mysticism cannot be underestimated. Science and spirituality are reconciled. For the new millennium this is shattering. Nothing will be the same again. All our beliefs about everything will change.

The dance of Shiva in Hindu mythology epitomizes all is energy. Revealing how energy forms matter, the vortex shows us how Eastern mysticism has anticipated modern scientific thought. Yoga is set to completely revolutionise our understanding of the Universe. Through the vortex we can appreciate how the dynamic in energy sets up the static in matter. It confirms what mystics of the East have always known, 'we live in a non-material world' and underpins the Vedic principle that the Universe is a non-material state of unending change and movement.

The vortex portrays particles of matter as particles of pure movement where nothing substantial exists that moves. These particles are more like thoughts than things. They are acts of consciousness, not acts of material substance. Quantum reality is more abstract than concrete. The universe appears as a mind and everything in the 'holographic world' we inhabit, including ourselves, seems to be an act of imagination. The Universe could be a dream and we are all dreamers within it. The new physics of consciousness confirms aboriginal belief that we exist in a dream state. The spiritual fits with the aboriginal belief that other dream states impact our dream world. In the parlance of the new science we would say the spiritual is another state of quantum reality.

It is impossible to prove the existence of other worlds spoken of in religion. However, the vortex explains the known world of matter in a way that unifies the whole of physics. The vortex account for physical reality is then extrapolated to provide a rationale for other realities. Suddenly the instrument of science provides a reasonable basis for religious belief and spiritual experience. An account appears in physics for unseen worlds that many people experience as true. When, through the lens of the subatomic vortex, we realise maya - the illusion of materialism - the barrier between science and spirituality melts away and they become one – a single body of knowledge in a multidimensional Universe where consciousness underlies everything.

Jeans James: *The Mysterious Universe,* Cambridge University Press, 1930

Chapter 1

The Dance of Shiva

Shiva is the god of creation and destruction who sustains through his dance the endless rhythm of the universe.

Fritjof Capra

The understanding that energy underlies everything is fundamental to Eastern mysticism.

The order of nature was conceived by the Vedic seers, not as static divine law, but as a dynamic principle which is inherent in the universe. This idea is not unlike the Chinese conception of the Tao – the way – as the way in which the universe works i.e. the order of nature. Like the Vedic seers, the Chinese sages saw the world in terms of flow and change, and thus gave the idea of a cosmic order an essentially dynamic connotation and Shiva, the Cosmic Dancer, is perhaps the most perfect personification of the dynamic universe.

Fritjof Capra

The schism between science and religion began in ancient Greece. Philosophers such as Democritus argued against the soul and beliefs of religion. Democritus speculated; *nothing exists except atoms and empty space. Everything else is opinion.* This, the atomic hypothesis, is the bedrock of modern science.

Democritus is considered the father of materialism.

14

His idea that material substance underlies everything is so fundamental to scientific thinking that we in the West are liable to accept it without question as truth.

Democritus

Democritus was introduced to his atomic hypothesis, around 500 B.C.E. by Leucippus who was of the opinion that matter is formed of irreducible particles, with predetermined properties. Democritus was the first to call the ultimate particle in matter 'atom' meaning that which cannot be cut. His concept was brought into the modern era by the writings of Epicurus (342 - 270 B.C.E.) and the Roman Lucretius (99 - 55 B.C.E.) in his poem 'De Rerum Natura'.

Newton was inspired by Lucretius when he wrote: *"It seems probable to me that God, in the beginning, formed matter in solid, massy, hard, impenetrable, moveable particles."*

The idea that everything was formed out of atoms moving in empty space is absolutely fundamental to modern scientific thought. Physicists imagine everything to be formed out particles in motion. They consider this to be the single most important principle in science.

Richard Feynman stated in one of his lectures: *"If in some cataclysm, all of scientific knowledge were to be destroyed, and only one sentence passed on to the next generation of creatures, what statement would contain the most information in the fewest words? I believe it is the atomic hypothesis... that all things are made of atoms — little particles that move around in perpetual motion."*

This statement by Richard Feynman summarises classical scientific materialism which still dominates the belief of many scientists today. Surprisingly there is no basis for the atomic hypothesis in physics. Materialism is a philosophy more akin to religion than science. Its tenets of faith are adhered to as vehemently by scientists as is the opposing religious thought by religious fundamentalists.

The atomic hypothesis gave rise to the classical model of the billiard-ball atom – the concept of an ultimate particle of matter with irreducible properties. Quark theory is a continuity of this classical thinking.

Quarks are conceived to be the ultimate particles in matter, possessing irreducible properties such as mass and fractional charge, charm and strangeness. If Newton were alive today, perhaps he would have said: *"It seems probable to me that in the big bang, matter formed mainly as massy, hard, up and down, top and bottom, strange, and charmed quarks."*

When the top-quark was supposedly discovered at Fermilab in April 1994, Jim Dawson opened his report in the 'Star Tribune' with the statement: *"So there we have it, after more than two thousand years of searching, all of the fundamental stuff of Democritus' atom has been revealed. The crowning moment came a couple of weeks ago, when physicists announced that a gigantic, 5,000-ton machine apparently had detected a very small particle called the top-quark."*

In fact they didn't see a quark; all they detected was a stream of particles that they took to be remnants of a quark. No quark, as such, has ever been isolated.

Quantum mechanics is a classical theory in so far as it embodies the atomic hypothesis assuming the existence of particles that move. Quantum mechanics should approach reality in terms of particles of motion rather than particles that move. This is because everything is formed out of particles of energy. Energy is the ability to do work. This means it is activity or potential activity. Particles of energy are particles of motion or potential motion. A quantum of energy is more a particle *of* motion than a particle *in* motion. In a sentence the quantum hypothesis should read: *All things are made of energy - little particles of perpetual motion.*

The issue of talking about particles of motion rather than particles in motion may sound slight but it is all-important. Understanding of the Universe hangs upon it.

There is motion but there are, ultimately, no moving objects; there is activity, but there are no actors; there are no dancers, there is only the dance.

Fritjof Capra (The Turning Point)

People in modern society are caught in the illusion of materialism, which has its roots in the philosophies of ancient Greece and Rome. The concept of material was never part of ancient Indian thought.

Buddha formulated a philosophy of change. He reduces substances, souls, monads, things to forces, movements, sequences and processes, and adopts a dynamic conception of realty.

S. Radhaksuhanan

Buddhists call the universe of ceaseless change

samsara, which means 'incessantly in motion'. Unlike materialists who believe in the existence of material particles that move, Buddhist thought denies the existence of anything that moves. There is nothing in the universe but change and motion, process and force.

Motion creates the illusion of material substance. This is Maya in Buddhist and Indian philosophy. This is why Buddhists affirm there is nothing in the world worth clinging to. To the Buddhist, an enlightened being is one who does not resist change, the endless flow of life, but moves with it.

Enlightenment is synonymous with surrender of resistance.

Scientists proclaim that classical physics is a thing of the past, but they are unwilling to surrender their attachment to it. Consequently the quantum concept has been turned into a classical theory. Physicists have done this by assuming the existence of particles that move to account for everything at a fundamental level and the uncertainty principle has served to shroud these particles with uncertainty so that nobody can be really certain what they are.

With Max Planck, Albert Einstein founded quantum theory. He was the true and original quantum theorist because he believed that matter and light were related to movement rather than material substance.

According to Einstein, the sole universal constant is the speed of light. Einstein contended that everything in the known universe is relative to this supreme speed of movement. At the same time he showed that mass could be equated with energy. By relating matter to energy, and also to the speed of light, he showed that energy is movement at the speed of light and that this movement is fundamental to matter.

Einstein never spoke about anything moving at the speed of light. He stood apart from the rest of science by accepting the existence of movement without assuming the existence of something that moved.

Einstein understood the non-material nature of the world in which we live. Few people really understood Einstein and this, according to William Berkson, was, "...*not because of his ideas or the mathematics he employed, but because of his worldview. Einstein denied the substantiality of matter and the field, whilst maintaining their reality.*"

Einstein stood alone as a non-classical theorist. He saw mass, space and time as being relative to the invariable speed of light. For him that was possible because he was able to take the tremendous leap of free thinking which allowed that movement could exist as the prime reality underlying, not only particles of light and matter but also the space in which they move.

The materialist cannot accept that everything is formed out of pure activity without any underlying substance; that nothing substantial really exists. Most people in materialistic society resist the idea that we are living in a non-material world.

When confronted by this stark reality on non-materiality in the definition of energy, Richard Feynman had the honesty to reply. "*It is important to realise that in physics today, we have no knowledge of what energy is.*"

The Eastern philosopher is not confronted be the same dilemma as the Western scientist because his fundamental assumptions are closer to the idea of everything in existence being an inherent dynamic state:

The general picture emerging from Hinduism is one of an organic, growing and rhythmical moving cosmos; of a universe in which everything is fluid and ever changing, all static forms being maya, that is, existing only as illu-

sory concepts. This last idea – the impermanence of all forms – is the starting point of Buddhism. The Buddha taught that 'all compounded things are impermanent', and that all suffering in the world arises from our trying to cling to fixed forms – objects, people or ideas – instead of accepting the world as it moves and changes. The dynamic worldview lies thus at the very root of Buddhism.

Fritjof Capra (The Tao of Physics)

In Hinduism, the creative energy of the Universe, the energy of endless transformation and change is represented as the 'Dance of Shiva'

The dancing Shiva is depicted in bronze sculpture with

four arms whose superbly balanced and yet dynamic gestures express the rhythm and unity of life. The upper right hand holds a drum to symbolise the primal sound of creation. The upper left carries a tongue of flame, the symbol of the element of destruction that brings about the end. The lower right hand is raised to sign us not to fear whilst the lower left points to the uplifted foot symbolizing our release from the illusion of maya. Shiva is depicted as dancing on the body of human ignorance, which has to be conquered before liberation can be attained. The face of Shiva is calm and detached as in the eye of the hurricane, He is at peace in the centre of the storm of change and transition He initiates, as worlds are destroyed and then recreated...The dance of Shiva symbolizes, not only the cosmic cycles of creation and destruction, but also the daily rhythm of birth and death, which is seen in Indian mysticism as the basis of all existence. At the same time Shiva reminds us that the manifold forms in the world are maya – not fundamental, but illusory and ever-changing – as he keeps creating and dissolving them in the ceaseless flow of his dance.

Fritjof Capra

His gestures wild and full of grace precipitate the cosmic illusion; his flying arms and legs and the swaying of his torso produce – indeed they are – the continuous creation-destruction of the universe, death exactly balancing birth, annihilation the end of every coming-forth.

Heinrich Zimmer

So how do the static forms of 'maya' or apparent materiality, arise from the dynamic state of energy? There are many manifestations of energy. Imagine picking up a piece of flint from the beach. Potential energy describes the energy stored by the flint. For example, whilst it is in your hand the flint stores energy that would be released should it fall and strike the ground. The potential energy is transformed into kinetic energy if you let the flint fall. The flint falls because of gravitational energy. As it strikes the shingle, radiant energy can be seen as in a spark and sound energy is heard. The flint gains heat energy in the warmth of your hand. Chemical energy was involved in forming the flint millions of years ago and it is electrical energy, which holds the flint in crystalline form through the continual interaction of electrons and protons in its atoms. At the same time however, the very substance of the flint itself can be reduced to energy. Physicists refer to mass energy when they talk about energy trapped in sub-atomic particles and nuclear energy when this trapped energy is released.

In classical physics, energy was defined as the ability to do work. It was understood as the ability of things to act and move. In modern physics this simplistic understanding of energy became confused by the discovery that nuclear energy could be derived from the destruction of mass. The classical definition of energy appeared to have collapsed. How it is possible to picture mass as a form of energy?

Energy can be stored in matter and the key to understanding this potential energy is to understand matter as a form of motion in which energy can be contained and harnessed. Energy in motion is called kinetic energy. Just as the potential energy in mass can be converted into kinetic energy in an atomic explosion, so in cosmic ray and high-energy research kinetic energy can be transformed back into mass.

To fully appreciate that mass is nothing but movement; imagine rain in the Swiss Alps. As the water falls to form streams tumbling down the mountains, these join the torrent of rivers that pass through hydroelectric plants. There the fall of the water is converted into the spin of the turbine and then the flow of electricity. The electricity is fed into CERN, the European particle accelerator where it is used to accelerate protons. As the protons collide in the intersecting rings of this high-energy laboratory, their motion is arrested. The arrested kinetic energy is transformed into the mass of a host of newly created particles.

Is anyone going to suggest that in the fall of rain, and the tumble of streams, the torrent of rivers and the spin of turbines, the flow of electricity and the acceleration of protons some mysterious material substance is transmitted to be transformed into new particles. As nothing was fed into the process apart from movement, common sense would suggest that the newly formed particles of matter are nothing but forms of movement.

The principle of *The Dance of Shiva*, that energy and therefore mass, is nothing but pure movement, is graphically illustrated by a Nobel Prize winning cosmic ray photograph. Just after World War II, Professor Cecil Powell, of Bristol University, developed a special, 1mm thick, 80% silver bromide photographic emulsion for detecting high-energy cosmic ray reactions and became renowned for littering the tops of Welsh mountains with stacks of photographic plates in black packets. Powell would leave them lying around for several weeks in the hope that a cosmic ray particle from the sun, penetrating the light proof wrapping, would collide with the nucleus of a heavy silver atom in the photographic emulsion. He later improved his techniques by sending his plates up in weather balloons. He needed altitude for his experiments because cosmic rays are more numerous on high ground than at sea level.

By analysing the tracks left in the developed photographic plates, Cecil Powell was able to study the results of high-energy sub-atomic reactions.

One photograph witnessed the collision of a cosmic ray alpha particle (nucleus of a helium atom) with the nucleus of a silver atom. In the collision the 3,000 billion electron volts of kinetic energy, carried by the alpha particle, was transformed into mass represented by a shower of 140 pi–mesons. These rapidly decayed, reverting to radiant energy.

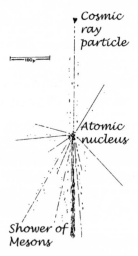

There is no experiment in the history of science to prove that matter is formed out of material substance whereas this cosmic ray experiment shows that mass is a form of movement. Physics has proved, beyond a shadow of doubt that we live in a non-material world in which nothing exists but the particles of activity we call energy. This is the allegoric *Dance of Shiva*.

The *Spin of Shakti* is an allegory for the way the dynamic state of energy forms the apparent static state of matter. The *Spin of Shakti* also accounts for potential energy, mass energy, and gravitational energy.

Matthews R, *Unraveling the Mind of God* Virgin Books 1992

Feynman Richard (with Leighton & Sands) *The Feynman Lectures on Physics* Addison Wesley, 1963

Dawson Jim, Star Tribune of Minneapolis - St. Paul USA, May 15th 1994

Zukav Gary, *The Dancing Wu Li Masters* Rider 1979

Capra Fritjof, *The Turning Point*, Fontana 1983

Capra Fritjof, *The Tao of Physics*, Wildwood House 1975

H. Zimmer, Myths and Symbols in Indian Art and Civilisation, Princetown University Press 1972

S. Radhaksuhanan, *Indian Philosophy,* Allen & Unwin 1951

Clerk R.W. *Einstein: His Life and Times* Hodder & Stoughton 1973

Berkson, William, *Fields of Force: World Views from Faraday to Einstein*, Rutledge & Kegan Paul, 1974

McKenzie A. E., *A Second MKS Course in Electricity*, Cambridge University Press, 1968 (Powell's cosmic ray photograph taken from plate 19)

Chapter 2

The Spin of Shakti

The vortex theory for matter is of a much more fundamental character than the ordinary solid particle theory.

J. J. Thomson

How is 'potential' energy contained in the static state of mass? This great mystery of $E=mc^2$, was solved thousands of years ago by mystics in the sub-continent of India. Yogis in ancient India discovered a form of energy that 20th century physics has overlooked. This is spin and it is the primordial spin of energy — which I describe allegorically as the Spin of Shakti — that completes the understanding of energy.

Indian mystics in the pre-scientific era were able to probe the atom with their minds. The first authoritative and systematic exposition of this practice by Yogis was written about 400 B.C.E. in the Sutras of Patanjali where the physiological and psychological results of meditation, concentration and contemplation are described in detail.

In Aphorism 3.26 it states: *pravrtty-aloka-nyasat suksma -vyavahita -viprakrsta -jnanam* e.g. knowledge of the small, the hidden and the distant can be acquired by directing the light of super-physical faculties. In Sanskrit such faculties are called 'Siddhis'. The Siddhi for perception of the small, the distant and the hidden is called *anima*. This siddhi is developed in the practice of advanced yoga.

In an altered state of consciousness the Yogi exercising

the siddhi can experience visual images of objects too small for human sight to discern. The experience is shrinking to a size commensurate with the objects being viewed. Through exercise of the anima siddhi yogis perceived that energy — *prana* — exists in matter — *akasa* - in the form of vortices — *vritta*.

The idea that spin is fundamental to matter appeared in the 19th. century but was lost in the 20th. A towering genius in late 19th. Century science was Lord Kelvin. Kelvin was the father of thermodynamics — the science of energy. In Kelvin's day the atom was considered to be the ultimate particle of matter, but whilst Kelvin believed in atoms, he had moved beyond the classical assumption that the smallest particles in matter occurred, like billiard balls, with a wide range of irreducible properties. The atomic hypothesis, embodied in the billiard ball model, was repugnant to him. Lord Kelvin found this model to be completely unsatisfactory because it offered no explanation for the most fundamental properties of particles of matter. He felt that the popular, materialistic view of matter was superficial and naive and he dismissed the billiard ball atom, as a 'monstrous assumption.'

In the Victorian era it was taken for granted that the Universe was pervaded by a frictionless ether which transmitted waves of light like the ocean transmits water-waves. Like most scientists, Kelvin believed that light consisted of wave motion in the ether. His addition to the theory was the idea that atoms were vortices in the ether.

Lord Kelvin suggested that the whole Universe could be reduced to two fundamental forms of motion, waves and vortices. This was not just a minor idea that might have been overlooked. It was developed into a major theory that dominated physics in the latter half of the 19th. Century and continued to be taught at Cambridge until

27

1910. James Clerk Maxwell, a major proponent for the vortex idea, wrote in the Encyclopaedia Britannica of 1875: *The vortex ring of Helmholtz imagined as the true form of the atom by Thomson (Lord Kelvin), satisfies more of the conditions than any atom hitherto imagined.*

Sir J. J. Thomson, who discovered the electron, said that the vortex theory for the smallest particles of matter..."*has a priori very strong recommendations in its favour*".

The vortex provided a fundamental understanding of matter in the late 19th. century. It was lost in the 20th. century but in the 21st. century a new vortex hypothesis could show how fundamental particles of matter are formed out of energy. The vortex could also show how energy is stored, in potential form, in the particles and forces associated with matter.

In his Special theory of Relativity, Einstein proclaimed that mass is equivalent to energy and that mass, space and time, are relative to the speed of light. For most of his life, Einstein searched in vain for a unified field. The vortex could be the key.

Kelvin believed that vortex motion created the properties characteristic of matter. In the vortex he endeavoured to reduce the properties of matter to a single, underlying and unifying principle. As such he was moving toward a unified field theory but his stroke of genius was to describe the underlying principle as a form of movement. It only required the suggestion that this was energy and that the speed of movement was the speed of light and the mystery of matter might have been solved. Kelvin was so close. He described light as a wave motion and matter as vortex motion in the same field. But he failed on three counts. Firstly he limited his vortex model to a vortex ring, depicted by a smoke ring when he could have allowed for other types of vortex motion. Then he thought the vortex particle was the atom and finally he imagined the field was the ether.

Toward the end of Kelvin's career there were many new breakthroughs in understanding the so-called atom, such as spectral lines, which his vortex model could not explain. It was still decades before the atom was to be split, so he had no reason to envisage sub-atomic vortices. Instead, he abandoned his model even though others continued to believe in it. Then, early in the 20th century, when the ether theory was dismissed, with it went most of the ether models, including Kelvin's vortex.

The vortex works, not when applied to the atom but to the smallest particles of matter — as originally intended by Kelvin. It is the subatomic particles, rather than the atoms themselves, which should be treated as vortices. The vortex idea is a casualty of the error in science of applying the term atom to the wrong thing — to a conglomeration of fundamental particles — protons, electrons and neutrons — rather than to the elementary particles themselves.

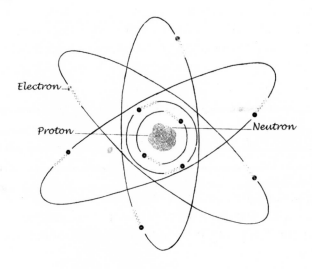

Electron

Proton

Neutron

29

Kelvin's vortex atom provides a historical precedent for the vortex in Western science. The outmoded vortex atom from classical physics was close to the truth. Worthwhile ideas survive the test of time. They are persistent. If ever they are lost they crop up again and again in history. The vortex is just such an idea.

The early atomists in ancient Greece, Democritus and Epicurus, believed that atoms formed bodies through vortex motion and thousands of years before them, mystics in ancient India considered vortex motion to be fundamental to matter.

From Einstein's Special theory of Relativity and Planck's Quantum theory it can be deduced that the activity called energy exists as the basis of everything, it is divided into particles and occurs as movement at the speed of light. These particles of activity are not objects moving, they are simply events at the speed of light. Even though a particle of activity could never be grasped as a substantial thing it should be possible to represent it with a clear and simple model.

The simplest model is a line. The line has no material substance. It is purely a representation of motion at the speed of light. The movement of light exists with direction therefore each particle of activity has a definite form based upon the direction that the movement takes within it.

In simplest terms waves are the direction of motion in particles of light and spin is the direction of motion in subatomic particles of matter. These two primordial forms of energy are represented allegorically as the 'wave dance' of Shiva and the 'vortex spin' of Shakti.

The model for the wave-form of energy — which is fundamental to light, heat, radio waves, gamma and X rays — is a line undulating in a bundle as a train of waves. The length of the line depicts the amount of energy in the bundle. This is proportional to the number of waves it contains, which explains why the energy in light is proportional to frequency.

Physicists have attempted to explain energy entirely in terms of its wave properties. I believe in the wave theory we have only half the picture. I am now convinced that physics is completed by the vortex as the second fundamental form of energy in nature. A vortex is a dynamic three-dimensional spiral. Most vortices are conical; the spin is about a single, central axis. However, Lord Kelvin's smoke ring was a vortex but it wasn't conical – and neither is a ball of string.

How is it possible for a 'corpuscular' elementary particle of matter to be a vortex? If the line of movement, at the speed of light, were to spin on a single point, on constantly changing axes, it would create a spherical vortex. This is depicted by a ball of wool.

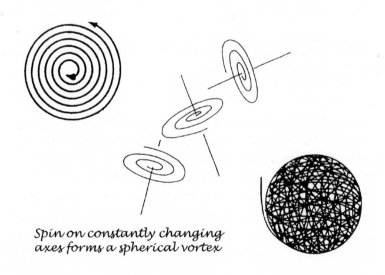

Spin on constantly changing axes forms a spherical vortex

A ball of wool or string has no measurable axial poles because the axes of spin, and therefore the poles, are changing all the time. A ball of wool represents the sub-atomic particle of matter as a spherical vortex.

A ball of wool is static, but a particle of energy spinning at the speed of light would be dynamic and as such it would be a vortex. Can the Universe be explained with something as simple as a ball of wool? Why not? That is the way it should be!. As Lord Rutherford said, *"These fundamental things have got to be simple."*

The winding of the wool on or off the ball represents the direction of spin of energy in the vortex. Appearing as a ball, the spherical vortex defines a volume and corresponds closely to the accepted three dimensional, corpuscular model for a sub-atomic particle of matter.

Einstein visualised matter as frozen light. In his famous equation, $E=mc^2$, he showed that vast amounts of energy are contained in minute amounts of matter. The 'wool-ball' model, by illustrating particles of matter as 'spinning light', explains the enormous energies released in nuclear explosions. Just as the ball of wool is a very compact form of yarn so the spherical vortex is a very compact form of energy. Unwind a ball of wool and you will have a roomful of twine.

The vortex, as a three dimensional spiral, suggests a continuous system of energy. Quantum theory contends that energy is discontinuous. Imagine an infinitely large ball of wool. At its centre it would tend to be a spiral but in the outer reaches it would tend to be concentric spheres — like a set of nested Russian dolls.

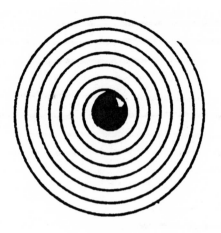

The concentric spheres of energy expanding or contracting establish the dynamic nature of the subatomic
vortex of energy. Imagine the set of dolls all growing at
the same rate. Alternatively imagine them all shrinking
together. This would represent the opposite directions of
flow of energy in the vortex. This simple idea provides an
account for electric fields.

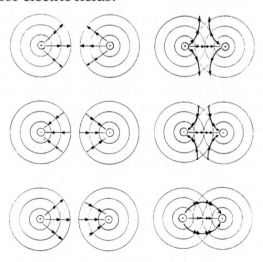

As the concentric spheres of energy in the vortex overlap they interact. The lines of electric force follow the points of contact between the spheres. If both sets of concentric spheres are expanding or contracting the lines of force set up repulsion between them. This is why like charges repel. If one set of concentric spheres is expanding and the other is contracting then the lines of force set up attraction. This is why opposite charges attract.

Energy is neither created nor destroyed and there appears to be no limit to the extension of concentric spheres of vortex energy. As the concentric spheres of energy expand, the intensity of energy in them may diminish but would never fall to zero. One could imagine the vortex energy being stretched and diminishing in intensity but never vanishing altogether much as a pie would never disappear by being dividing it into ever-smaller pieces.

The infinite extension of the vortex can account for the unlimited range of electric charge, magnetism and space. Because of its unlimited extension the concept of size cannot be applied to the vortex. One vortex particle can only be considered to be greater or lesser than another in terms of its mass-energy. The proton is a vortex containing nearly two thousand times as much mass-energy as is contained in the electron. It is this mass-energy difference that is represented by the relative sizes of the vortices. It is better to describe vortices in terms of inertia than in terms of size and the vortex explains inertia.

In 'The Character of Physical Law' Richard Feynman said *"The law of inertia has no known origin."*

Thanks to the Yogic anima siddhi an account for inertia is forthcoming.

Energy is motion. Motion creates inertia. This principle is illustrated when you ride a bike or skis. The faster you move the more inertia you possess and therefore the easier it is to balance.

34

The flow of energy in a train of waves creates a particle of kinetic inertia and the flow of energy in a vortex creates a particle of static inertia. Kinetic inertia is the tendency of something to keep moving unless it is stopped and static inertia is the tendency of something to stay put unless it is moved.

Because of spin, static inertia is a feature of a spherical vortex. If you take a gyroscope, the spin sets up inertia in the plane of the spin. A spinning pebble skipping across the surface of a pond also illustrates this point. The spin of the pebble, keeps it in the plane in which it is thrown so that whilst it is spinning it skims across the surface of the water rather than sinking.

In a spherical vortex the simultaneous flow of energy in all directions would set up an inertia or resistance to movement of the vortex in any direction. It is this static inertia that we perceive as mass. Mass is created by vortex motion. Mass quantifies the inertia of the vortex.

Photons of light do not possess mass because they are based on wave not vortex motion. It is the vortex form of energy, not energy itself that creates mass. Mass is created by vortex motion so mass, momentum and angular momentum are a properties of the vortex, not the energy within the vortex

Vortex motion sets up *maya*; the illusion of material. Democritus was deluded by the vortex motion of energy into believing that fundamental particles are real and solid. This error in the *atomic hypothesis* of Democritus led classical Western Science to support the illusion we know as materialism.

Clerk-Maxwell James *Scientific Papers* vol ii, pp 445-84, 1890

Coomaraswamy, A. K., *The Dance of Shiva*, The Noonday Press 1969

Feynman Richard, *The Character of Physical Law,* BBC Publications, 1965

Prabhavananda S & Isherwood C, *Translations of the Yoga Aphorisms of Patanjali,* Allen & Unwin 1953

Ramacharaka Yogi, *An Advanced Course in Yogi Philosophy* Fowler

Radhaksuhanan S, Indian Philosophy, Allen & Unwin, 1951

Thomson J. J., *Treatise on the Motion of Vortex Rings,* Cambridge University. 1884

Thomson, William: *Proceedings of the Royal Society of Edinburgh,* vol vi, pp. 94-105 (reprinted in *Phil Mag.,* vol xxxiv, pp 15-24, 1867

Thomson, William: *Mathematical and Physical Papers* 6 vols. 1841-1882

Thomson, William; *Popular Lectures and Addresses* vol i

Thomson, S.P., *Life of William Thomson, Baron Kelvin of Largs* 1910

Chapter 3

The Myth of Materialism

"Right understanding is part of enlightenment."

Buddha

The myth of scientific materialism was exploded early in the 20th. century by Einstein's equation: $E=mc^2$ but people find it hard to imagine how mass could be pure energy. The great value of the sub-atomic vortex is that it provides a clear picture of how movement forms mass.

Particles of vortex energy and their movement within the atom set up the illusion of material substance in mass. Materialistic philosophers define a material thing as something possessing mass and inertia. It extends in three dimensions and has a capacity for kinetic and potential energy. Material things are known to act on each other at a distance through a number of forces and they can be transformed into energy.

The concept of material substance originated in ancient Greece where philosophers relied on speculation and reason. They didn't do experiments or deploy the direct observational powers ancient Indian philosophers used in their investigations. The Greek philosophers were taken in by the illusion of vortex motion.

Vortex energy in the atom accounts for potential energy, mass, charge, magnetism, gravity, the strong nuclear force, the inertia of matter, the extension of matter in three dimensions and mass-energy transformations.

The properties of material things can be explained away by vortex motion.

Western scientists probed the atom and used the knowledge they acquired to invent of weapons of mass destruction. Eastern mystics looked into the atom and, unperturbed by uncertainty, used the knowledge they acquired, in poetic and mythical images, to uplift the human spirit.

The Hindu mystics saw physical reality, not as static material, but as a dynamic principle fundamental to the universe. In the Bhagavad-Gita Krishna says, *"If I did not engage in action, these worlds would perish"*

The Chinese Tao – the way – epitomises the dynamic way in which the universe works. Like the Vedic seers, the Chinese sages saw the world in terms of flow and change, viewing the cosmos as intrinsically dynamic.

The vortex is the key to understanding how the dynamic can appear static. The vortex of energy creates the illusion of rest. Movement in a straight line represents maximum velocity. As the line of movement, representing energy, follows a curved path the velocity in the straight line is reduced.

If the line of movement is directed in a spiral, the velocity in the straight line is zero. Thus the vortex is the form in which energy can appear to 'stand still'. It is this movement of energy in three dimensions into or out of a single point in the spherical vortex, which creates the 'stay-put' inertia characteristic of matter.

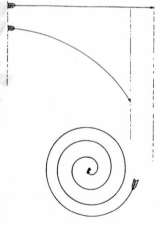

The greatest challenge for the Western mind is to conceive of movement underlying everything when nothing substantial exists to move. How would you define the material substance of the world?

If you say, "Matter is long lasting so material things are durable and stable," the reply would be, "the vortex is both durable and stable".

You might retort "Material things are solid; they stay put and resist change."

The counter argument would be, "Things are the way they appear to be. Materiality is an illusion created by the vortex. As energy spins to form matter it sets up inertia and mass. Spin creates the impression of solidity in matter".

You could cite extension in three dimensions as a property of material substance but then extension in three dimensions is a property conferred upon matter by the spherical vortex of energy.

If you wonder how energy traveling at the speed of light can make a stationary particle of matter the answer would be that when light follows a curved trajectory its velocity in the straight line is reduced. The velocity of light in space is slightly less than the top speed of light because space is curved. Curved space slows light down! If light spins into a single stationary point its linear velocity would be zero. When the movement of light is in spin it is still energy only it isn't propagating. It's not going anywhere, so it stands still; it is the equivalent of a standing wave. The vortex is a standing form of energy that creates the illusion of solidity.

You might argue that material is something you can see and feel. It has shape and size and fills space. My response would be, "You see matter only because photons of light interact with the electro-magnetic fields of orbiting electrons in atoms. Some photons are absorbed and others reflected depending on their frequency.

"As white light hits the atoms in a leaf all frequencies except for green are absorbed and used in photosynthesis. Only the frequency of green light is not used to lock energy into sugar molecules. This rejected frequency bounces back out of the leaf or passes right through it and enters your eyes enabling you to see the leaf as something green.

"When you feel something you don't actually touch it. Electrons in the atoms of your hand repel the electrons in the object so no real contact occurs. All seeing and feeling can be explained away as electro-magnetic interactions.

"The electric bonds between air molecules are so weak, the nitrogen and oxygen slip around us as we move through them. When we swim the bonds between hydrogen atoms in one molecule of water and the oxygen atom in another are much stronger. The links are called hydrogen bonding.

"You wouldn't dive into a swimming pool full of air but you would dive into a swimming pool full of water because the hydrogen bonds are strong enough to break your fall. If you dive into a pool full of air instead of water, the air wouldn't break your fall. Instead you would very likely break your neck on the bottom of the pool because of the strong electric bonds between calcium, silicon and oxygen in the tiles, cement and concrete at the base. They don't give way at all.

"The rigidity of a solid is the strength of electro-magnetic links between atoms. The difference between solids, liquids and gasses is just a difference in strength of atomic bonding. All the properties of atomic and molecular matter can be explained in terms of the electro-magnetic interactions of vortices between atoms."

The vortex provides a simple account for electric, magnetic and gravitational forces. The ability of one particle of matter to act on another at a distance is explained by vortices overlapping and interacting. If they were inert blobs of material they could not act.

It is not possible for subatomic particles to give and receive energy for electric attraction or repulsion. The sun continually gives out energy but it is a vast body with enormous reserves of nuclear fuel. It is impossible to imagine a minute particle having the infinite reserves of energy needed to account for the force of electric charge that extends into infinity.

Imagining the subatomic particle as a form of movement overcomes this problem. Electric charge is an intrinsic property of the particle itself, an expression of its innate dynamic nature. If particles are systems of energy rather than inert blobs of material then electric charge is easy to explain. Forces are not caused by energies associated with vortex particles, forces are an intrinsic property of the particles themselves

The vortex is the key to understanding the force fields associated with matter. As whirlpools of energy it would be completely natural for subatomic particles to act on each other. Take, for example, the electric forces that hold particles in an atom.

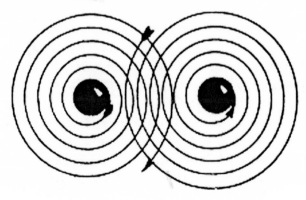

The spin in the subatomic vortex in or out of the centre - corresponding to the Chinese Yin and Yang - accounts for positive and negative charge.

When one vortex approaches another their extending vortex energy would overlap and interact long before the particle centres come together. These extended vortex interactions account for 'action at a distance'.

Coulomb's Law for charge states that the force of interaction between charged particles is inversely proportional to the square of the distance between them.

This is the inverse square law. This law dictates that as the distance from the centre of the vortex doubles, the intensity of energy drops to a quarter of its original value; when the distance trebles, the intensity of energy drops to a ninth, etc.

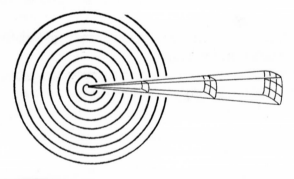

The strength of interaction between vortices is proportional to the intensity of energy overlapping. Because vortices are three dimensional extensions their interactions would obey the inverse square law. Charles Coulomb observed this for electric charges.

As the intensity of vortex energy falls away according to the inverse square law it would rapidly become negligible — so the particle of matter would correspond to the dense centre of the vortex. Energy is neither created nor destroyed therefore there is no end to the vortex. There is no real boundary or surface to the particle. The vortex of energy is an infinite extension.

If vortex energy extends out in all directions into infinity we could not be directly aware of the sparse vortex energy. It would exist beyond the bounds of our perception. We could only be aware of its existence through its effects such as magnetism, gravity or electric charge. This would be dark matter.

The surface of a particle is like the horizon. It does not exist in nature. It is merely the last of matter that we perceive. It is not the end of matter itself. To imagine that matter is confined to its apparent surface is the equivalent of flat earth mentality.

It is blindness to assume nothing exists beyond direct perception. The idea that particles of matter are bounded bits of material leads to the assumption that other things exist in and around particles to account for the way they act on each other. This is how the speculation arose that forces are the action of other things, apart from the particle, such as space-time curvature, virtual photons or force carrying particles.

Einstein imagined that matter distorts space-time. Newton assumed something was mediating between bodies of matter to cause them to act on each other at a distance. Today physicists suggest that force-carrying particles pop off subatomic particles like fleas off a dog.

If mass is a form of energy then particles of matter must be intrinsically dynamic. If energy is neither created nor destroyed these dynamic properties will be eternal and extend into infinity. If we dump materialism and embrace energy as the real basis of mass we can appreciate force as an intrinsic property of an energetic particle. The vortex enables us to visualise particles as energetic, extending, interacting swirling whirlpools of light rather than bounded blobs of lifeless material that bounce off each other.

Through the vortex we can make a paradigm shift. Fundamental assumptions and firmly held beliefs limit us in science as well as religion. The scientific world is stuck with an outmoded paradigm, the concept of material substance.

Materialistic thinkers use science to explain away religious phenomena and non-material worlds. The vortex explains away materiality and shows we are living in a non-material world!

Having accepted matter as a form of energy it is tempting to imagine energy as the stuff moving in the vortex. However, energy is not some stuff. Energy is pure activity. Energy is not something acting. Energy is the action. This is clear in the mystical traditions of the East. The Universe is not Shiva dancing, it is not something dancing; the Universe is the dance of Shiva, it is the dance itself.

It is important we appreciate that our world is formed not out of particles in motion but particles of motion. Replacing the word *of* with the word *in* between the words *particle* and *motion* had the power to totally alter our worldview.

Our current worldview originated in ancient Greece. Aristotle speculated that everything was formed out of four fluid mediums, *earth, air, fire* and *water*. We are all formed of air and water with minerals from the earth, and we are powered by the fire of the sun, captured by plants in photosynthesis. But this idea is a form of materialism because Aristotle assumed earth, air, fire and water existed first and then combined to form everything else. Any idea that something first exists and then moves is materialism.

44

The Old Testament of the Bible begins with a definitive statement of materialism in the first verse of Genesis: *In the beginning God created the heavens and the earth.*

Whether the primordial entity is called 'God' or 'atom', or if the emptiness occupied is called 'space' or 'void' is inconsequential. What matters is the underlying philosophy. Western religion and science are both founded on materialism.

In the New Testament John's gospel opens with the verse: *In the beginning was the Word and the Word was with God and the Word was God.*

Word is sound and sound is a form of energy therefore John's opening verse could be rewritten: *In the beginning was energy and the energy was with God and the energy was God.*

In that extraordinary statement John effectively declared energy to be the basis of everything. He also implied there is no separation between energy and the source of energy, no separation between God and the Universe.

St. John was influenced in his idea of the logos or word from the teaching of Hermes. The name Hermes in the Greek language and 'Mercury' in Latin both infer fluidity. In the 'Corpus Hermeticum' Hermes concluded: *Everything is movement.*

Consider the definition in science for energy. The word energy means 'the movement within'. If everything is formed of energy *then everything is formed of 'the movement within'.* This means that *within everything there is nothing but movement.* The key to understanding this conundrum is the vortex. The vortex is a system of movement that appears to be inert. It is a dynamic state that appears to be at rest. It is a state of constant change that resists change.

According to St. Augustine, Hermes Trismegistus was a real person who lived and taught in Egypt a few generations after Moses. Hermes warned, *"Base your reasoning on revelation not speculation."* Ancient Greek philosophers elevated Hermes to the status of a god but ignored his diction. They based their reasoning on speculation rather than revelation. Leucippus and Democritus speculated the atomic hypothesis. The vortex hypothesis, which reveals the myth in materialism, is based on the anima siddhi revelation.

We live in a paradoxical reality where mercurial vortex motion creates the illusion of materiality. Leucippus and Democritus were deluded by vortex motion. Vortex interactions set up atoms and molecules, crystals and rocks, lakes, trees and mountains. The three states of matter solids, liquids and gases can be traced back to vortex interactions.

It is energy in vortex motion that resists movement and change in matter. It is the extension of vortex energy beyond our direct perception that gives rise to the illusion of the void. The ancient Greek philosophers were steeped in naïve realism and deluded by their senses. Though Einstein vanquished materialism the myth is still accepted as the bedrock of science because the underlying philosophy of science was founded on the soulless speculations of ancient Greek philosophers.

The atheism of science, stemming from the myth of materialism, was an important step in the emancipation of Western society from religious fear and superstition. The power of the medieval church had to be broken. That was achieved by the renaissance and the reformation. However, there is a resurgence of religious faith as ever more people search for meaning in life. Because of disillusionment caused by religious sectarianism this search is occurring increasingly outside of the established religions. Within and without the religious establishment, spiritually orientated people are seeking a place for God in science. The time has come for science to recover its soul.

Ash David & Hewitt Peter: *The Vortex: Key to Future Science* Gateway Books 1990

Ash David: *The New Science of the Spirit* College of Psychic Studies, 1995

Darwin Charles: *The Origin of Species* 1859 (reprinted by Penguin Books)

St. Augustine, Civitas Dei XVIII. 29

Chapter 4

God in Science

This world that we see is the imagination of God

Maharaji

The essence of materialism is the assumption that the particle pre-exists movement. Modern physics has proved the reverse is true. Movement pre-exists the particle. E=mc² shows that mass is energy and energy is activity. Particles of energy are particles of pure movement. They are bits of activity, and activity in the form of the spherical vortex creates the massive, corpuscular subatomic particle. The vortex sets up 'potential energy' and the vortex deludes us into the naïve realism of materialism, into thinking something pre-exists or underlies energetic movement.

The activity we call energy exists but there is no thing or substance underlying this activity. Activity where nothing exists to act is abstract. In essence our world is abstract. Particles of light and matter appear to be more thoughts than things. The atom and quantum are more the nature of dream-stuff than concrete material substance.

If matter is non-substantial and we are living in a dream, humanity and the whole of history would be the equivalent of a hologram. The evolution of the Earth over billions of years, indeed the unrolling of the entire Uni-

48

verse might be an unfolding vision; nothing more than an incredible act of imagination. It would seem we are in the Matrix!

If the Universe is a vision, who or what is the visionary? If our world is a dream who is the dreamer? If you believe the universal dreamer is God then it could be said: *This world we see is the imagination of God.*

When in my teens I first pondered the possibility that the Universe is a dream I supposed God to be the dreamer. Realising particles are more thoughts than things I surmised the Universe to be a mind, the mind of God.

This idea has a precedent in Hinduism where Shiva is imagined as the source of the Universe and Shakti as the energy that is the Universe. The male-female duality of Shiva – Shakti fits with the male 'inseminating' and the female 'gestating' the body of creation. Shiva as the wave is the idea, the transmitting of information, Shakti as the vortex is the form, the manifestation of information. Shiva would be the dreamer; Shakti the dream.

As a Catholic boy I had been taught that God created everything out of nothing and compared to God everything was nothing. This concept now made sense to me. If the Universe is the imagination of God – abstract rather than material – then in essence it is as nothing.

In my teens I considered the number of atoms in a grain of sand and the number of grains in the deserts and seashores of the Earth. Then I meditated on the trillions of planets and stars in our galaxy and thought about the billions of galaxies. Each subatomic particle and photon of light is an individual act of imagination sustained in excess of a billion, trillion, trillion years. I found it hard to conceive a person capable of imagining such an ongoing state of creation. I also had problems with the word

'God' which came from old English for 'good'. I was searching for a word that would provide a deeper understanding of the major principle underlying the Universe. To begin with I followed Einstein's footsteps in treating God as a cosmic principle rather than a person. Then I realised God is a person, but not a person separate from me. God is the person in us all. I came to that realisation through meditation.

I had embarked on meditation because theoretical physics involved me in a lot of mental activity. Meditation quieted my mind and brought me inner peace. In meditation I connected with a centre of stillness where I experienced myself as pure consciousness.

When I reached real stillness there was no separation; I became the consciousness. Through meditation I had been shown my source of awareness as a pool, presenting a reflection of my true self. My thoughts were like a breeze that ruffled the surface, obscuring the reflection of tranquillity. Through meditation I realised conscious awareness to be more fundamental than thought. The real me was not a body of thought but the awareness of thought.

Descartes said *"I think therefore I am."*

Obviously Descartes didn't meditate.

Through meditation my experience is, *"I am therefore I think."*

Socrates said *"Know thyself."*

I realised from the Hermetic principle, as above so below, as below so above, through knowledge of myself I could come to know the Universe. Consciousness awareness is more fundamental than thought and this could apply to the Universe in macrocosm as to me in microcosm.

It dawned on me that maybe the principle we call God

is a universal conscious awareness that holds every particle of energy in existence as a thought and that particles
of energy are memories in the mind of this 'Universal
Consciousness'.

As I pondered on this I realised, whereas every human
is unique, as is every snowflake and no blade of grass is
the same, every proton is absolutely identical in physical
characteristics. All protons have the same measure of
mass and electric charge. They all have the same value of
quantum spin and the same estimated lifespan of a billion, trillion, trillion years. This suggested to me that the
consciousness awareness underlying them all is one and
the same. If it were a different consciousness behind each
proton, they would all vary in their characteristics.

The oneness of consciousness has profound implications. We are all conscious. If consciousness is indivisible
then we are all the same, one, undivided conscious
awareness looking out of six billion pairs of eyes, hearing
through twelve million ears feeling through a multitude
of bodies. Our thoughts may be different and our personalities unique but our emotions are not dissimilar. Each
of us lives an individual experience but in our beingness
we are the same. We are one being in many bodies.

There are no free lunches in the Universe. We receive a
body at birth but it is not free. We pay for human life in
experience. Whether good, bad or indifferent, the experiences gained in human life are invaluable to the Universe. The collective God that we are grows in wisdom
and understanding through the totality of experience
gained over millennia through billions of human lives.

In the state of oneness, consciousness cannot be
self-conscious. To be conscious of itself separation is necessary. This is impossible in reality but possible in the un-
real state of the abstract. By animating bodies in the

abstract Universe of energy, the one can experience an illusion of separation into the many. Through this illusion it can come to know itself through a multitude of forms and experiences. This is the 'human situation'.

Hermes understood this. That is why he called us human beings. The word 'hu' denoted God in ancient Egyptian. We are god-man beings — one in our divinity, diverse in our mortality.

I said "you are gods", and all of you are children of the Most High.

Psalm 82:6

Is it not written in your law, 'I said you are gods?

John 10:34

When Sai Baba was asked if he was God he replied:
"Yes but so are you; the difference is I know it and you don't."

This was a very profound statement. If we are God all we have to do is know it and we are free.

And the truth shall set you free. John 8:32

If we are God and we believe we are not, then by that thought we are enslaved. We may want to escape from the limited state of physical embodiment but freedom comes from inner knowledge not outer circumstances. We inhabit bodies to appreciate the state of limitation and unique experience the human situation provides. For God, being human is vital for self-knowledge. For humans, self-knowledge is vital for reunion with the self as God.

Great spiritual teachers, throughout the ages, have stressed the importance of self-knowledge beyond the ceaseless chatter of the mind. The Universe, as the manifest Word, is endless chatter. The Source of all is the silent awareness seated in the stillness underlying the eternal motion of energy. Within each of us is the source of all and this is our own essential being. Knowing 'who we really are' is to know the origin of all. This knowledge is our birthright. It is always there for us because it is our 'I Am' self. All we have to do is go into stillness to know it.

Be still and know that I Am God. Psalm 46:10

Though we may differ in personality we are one person in our essential being and that person is God. Thus it was in coming into the knowledge of my true self, beyond the stream of my thoughts I became aware of the intimate person known to all who seek within themselves; I came to know my true self as God.

"If then you do not make yourself equal to God, you cannot apprehend God; for like is known by like. Leap clear of all that is corporeal, and make yourself grow to a like expanse with that greatness which is beyond all measure; rise up above all time and become eternal; then you will apprehend God. Think that for you too nothing is impossible; deem that you too are immortal, and that you are able to grasp all things in your thought, to know every craft and every science; find yourself home in the haunts of every living creature… but if you shut up your soul in your body, and abase yourself, and say, 'I know nothing, I can do nothing, I am afraid of earth and sea, I cannot mount to Heaven; I do not know what I was, nor what I shall be'; then what have you to do with God? Your thought can grasp nothing beautiful and good, if you cleave to the body and are evil. For it is the height of evil not to know God; but to be capable of knowing God, and to wish and hope you know him, is the road that leads straight to the good; and that is the easy road to travel…for there is nothing that is not God. And you say 'God is invisible?' Speak not so. Who is more manifest than God?"

Hermes Trismegistos

The idealist George Berkeley (1685-1753) was of the opinion that everything existing is either a mind or depends for its existence upon a mind. He was described

53

appropriately as an immaterialist. Movement rather than material was the basis of his reality. As far as he was concerned matter did not exist. He adopted the position that ordinary physical objects are composed solely of ideas, which are inherently mental. Berkley was renowned for his motto, *'esse is percipi'* - to be is to be perceived. Parallels can be drawn between Bishop Berkley's thinking and the Copenhagen Interpretation of quantum reality that phenomena exist only to the extent that they are observed.

Berkley's idealism coincided closely with the Hermetic principle that *everything is mind*. I read of Berkley and his philosophy and the principles of Hermes years after developing my own ideas and was encouraged by the parallels in thought. At the time I was endeavouring to find a place for God in science I was unaware of my declaration as a four year old, that I would prove the existence of God through science. It was decades later that my parents told me what I had said in infancy.

When, in my teens, I first grappled with the concept of God all I had was an inner core of certainty that the ultimate truth once revealed would unite science and religion into a single stream of knowledge. When others said that the chasm between spirituality and science could not be bridged I was heartened by the words of my father:

"If you think something can be done then go ahead and do it."

Decades later I found in Corpus Hermeticum, the dictum: *Reason wills everything into existence.*

This principle of Hermes Trismegistos set me on another train of thought. Perhaps God is not an entity separate from the Universe but is pure reason within the Universe. Maybe God is the innate intelligence of the Universal mind?

The way Hermes used the word *Reason* as synonymous with God intrigued me. He also used the term Logos meaning 'the primordial word' which echoed millennia later in the Gospel of St John.

In the beginning was the Word and the Word was with God and the Word was God. He was in the beginning with God. All things were made through Him, and without Him nothing was made that was made.

John 1: 1-3

It was clear to me that Hermes and John were referring to the innate intelligence of energy. When I interpreted the 'Word' as energy the synonymy between God and the Word that wills everything into existence suggested to me there is no separation between the universal consciousness and energy.

The concept of God as an entity existing with predetermined properties is the same logic that contends material substance exists with predetermined properties. Energy is activity and consciousness is awareness. These are properties rather than entities, the equivalent of verbs rather than nouns. We know consciousness exists but if we assume the existence of an entity that is conscious we fall back into the trap of materialism.

Believing in a 'God' that is conscious is no different from Victorian scientists assuming the existence of an ether pervading space to carry waves of light. How can there be waves without an ocean? How can there be universal consciousness without a being that is conscious? It is as hard to envision pure unsupported consciousness as it is to imagine pure unsupported movement that we call energy. I am not saying there is no God. **I am suggesting God is more the state of pure consciousness than a being that is conscious.**

Energy is action. If it is not an act of material substance it can exist as an act of consciousness. If no thing exists that is aware, then consciousness exists only to the extent that it is aware; that it is consciousness of something.

If there is no pre-existing substance underlying energy and no pre-existing being underlying consciousness then each exists only by virtue of the other. Universal consciousness would exist because it is conscious of the thought forms we describe as energy. These acts of abstraction, in their turn, depend on conscious awareness to maintain their existence.

Returning to the analogy of the dream and the dreamer, it is easy to imagine God as a being dreaming the Universe. When the Universal dreamer wakes up the Universe then ceases to be. But that implies separation between God and the Universe. If, however, the dreamer is purely 'ability to dream' then it would depend on the dream in order to exist. Should the non-substantial dream vanish then the non-substantial dreamer would also cease to be. If this line of reasoning is correct consciousness and energy; God and the Universe, would be inextricably bound together. Neither one could exist without the other. This is clear in the opening passage of John: *In the beginning was the Word and the Word was with God and the Word was God.*

If space and time depend on the vortex then spatial and temporal separation between God and the Universe is impossible. If space and time are derived from the vortex these dimensions belong to a form of energy within the Universe. This means there cannot be a beginning or end to the Universe or to Consciousness. To both God and energy the same rule applies: neither is created; neither is destroyed. In Hindu mythology Shiva and Shakti are inseparable.

Maharaji Guru, Bihar Satsang, Patna, India 25th Dec. 1971

Churton Tobias, The Gnostics, Weidenfeld & Nicolson 1987

Graves Robert, Greek Myths, QPD 1993

Hermes Trismegistos, Corpus Hermeticum; XI.2

Chapter 5

Love thy Neighbour

Mysticism is the sower of true science

Paul Roland

Most people assume that space is the emptiness left behind when matter is removed, but Einstein, thinking differently from most people, believed that space was not an absolute void. He treated it as something real and connected to matter. The importance of this thinking in regard to relativity is clear from a cryptic remark he made when he arrived in New York, in 1919. When asked by a reporter to express his theory of relativity in a single sentence. He replied: *"Remove matter from the Universe and you also remove space-time."*

Einstein's original thinking about space began when he was only five. He was recovering from an illness and his father gave him a compass to play with. As he turned the compass, the needle kept pointing in the same direction. It struck the young Albert that space must be holding the needle. Playing with the compass, Einstein conceived of the idea that space wasn't nothingness, but was something as real as matter itself. It was this thinking that led him to his theories of relativity.

The vortex enables us to understand Einstein's relativity by providing a clear picture of space as a real form of energy. As energy in the vortex extends from the centre of a subatomic particle outward, its intensity would rapidly diminish to infinitesimal levels but would never reach zero. This is because zero is approached but never reached by dividing something into ever-smaller fractions.

Imagine you had a magic balloon that never burst. If you kept blowing it up, the rubber would become infinitesimally thin and the balloon would become infinitely large but the total amount of rubber would always remain the same — it would simply be stretched over a larger and larger area. In the same way, vortex energy would not vanish into nothingness, even though it extends out into infinity. It follows from this logic that every vortex must be as big as the physical Universe; that the vortex must be space itself.

If the infinite extension of vortex energy were space the connection between space and matter would make sense. This idea is supported by the fact that both matter and space are three-dimensional extensions. Space, like matter extends in three dimensions rather than two or four or more. The vortex — as a three dimensional system of energy — accounts for the three-dimensional extension characteristic of space.

The statement: *"Remove matter from the Universe and you also remove space-time"* suggests that Einstein saw space as something connected to matter. With the vortex model, it is possible to go a step further and suggest that space and matter are the same.

If matter were vortex energy we perceive and space were vortex energy extending beyond the limits of our perception — which is why it appears to be a void - there would be no real difference between matter and space.

If the vortex is responsible for both matter and space, as matter is divided into particles then space must also be discontinuous. The dense centre of the vortex would constitute the particle of matter and the sparse peripheral regions of the vortex would constitute a part of space. If a vortex were destroyed then a particle of matter and a part of space would disappear from existence. Thus, we can understand Einstein's statement: *"Remove matter from the Universe and you also remove space-time."*

Indoctrinated by the myth of materialism we have been led to believe that there is more space in an empty room than a room full of people. If Einstein is to be believed the reverse would be true. Each person would bring a packet of space with them when they enter the room.

The space in a room full of people is denser than in an empty room. Do we not experience this? The space in a room full of people feels thick. In the empty room after they have all left it feels thin.

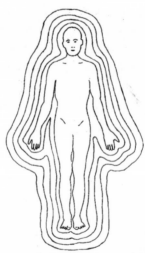

When we set up vibrations in our extension of space by thought and intent, our frequencies extend throughout the Universe touching everyone. They have a subtle influence on everything. Thus it is the collective thoughts of humanity affect the Universe. For example, war zones respond to the vibrations we transmit in our extending fields of space. Trouble spots in the world are influenced by whether we choose to transmit peace or anger, love or hatred in the course of our daily lives.

Bodies of matter can be visualised as surrounded by extending bubbles of space. These would occur as an extension of the shape of the body.

The concentric bubbles of space would extend from the body into infinity — each bubble fading away in intensity as it increases in size. At a distance from the body the intensity of energy contributed by its own bubble of space would be negligible. However, the summation of the bubbles arising from neighbouring particles would step up the intensity of space in that region. Space could be pictured as foam made up from the addition of all the bubbles of space reaching out from every particle and body of matter in existence. Concentric bubbles of space, extending from the Earth, the Sun and other planets and stars, would add up to form universal space.

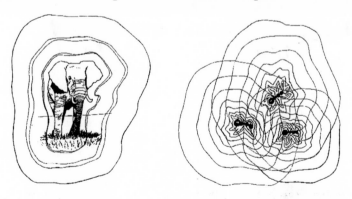

In the *Flower of Life* symbol we see the interlacing bubbles of space extending from a multitude of vortex particles. Drunvalo Melchizedek contends that this pattern is the most significant in science as it symbolises the interacting particles and forces of nature. His claim makes sense if the *Flower of Life* depicts the particulate nature of space.

Space bubbles would share, with elementary particles of matter, all the properties of the vortex. In fact, forces such as gravity, electric charge and magnetism would be more the property of space than the particles of which they are an extension, because they are effects of the vortex extending into infinity.

Space and gravity share identical properties:

Space	Gravity
1. Extends from matter.	1. Extends from matter.
2. Extends from the centre of each particle of matter.	2. Acts from the centre of each particle of matter.
3. Unlimited range in three dimensions.	3. Unlimited range in three dimensions.
4. Obeys the Inverse Square Law.	4. Obeys the Inverse Square Law.
5. Curves the path of light.	5. Curves the path of light.
6. Universal space is the sum of the components of space extending from every particle of matter.	6. Universal gravity is the sum of the components of gravity acting from every particle of matter.

Space is active electrically and only appears to be electrically neutral because there are an equal number of opposite charges in existence. The electric force is set up by the 'yin-yang' spin out or into the vortex of energy. Magnetism is set up by the rotation of the vortex.

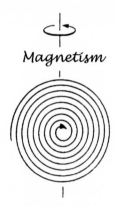

Magnetism

This is demonstrated by the natural magnetism of an electron caused by rotation of its electric field. In physics this is called its *magnetic moment*.

Space is magnetically active. The Earth's magnetic field is a property of the space extending from the Earth. The five-year-old Einstein was right in assuming it was space that influenced the compass needle. The assumption that space is a void occupied by magnetic forces is just another myth that people have come to believe. Magnetism appears to be a force field occupying space because one bubble of space occupies another. Magnetism is an effect of space.

In his special theory of relativity, Einstein predicted that space and time are relative to movement at the speed of light — which sets up a riddle: *How can space and time be relative to a speed of movement when speed is itself a relationship between space and time?* It is a chicken and egg dilemma, what came first, space-time or movement? This riddle of relativity can be resolved easily: *Each vortex as a particle of space and matter exists and moves in the space created by another vortex.*

Each vortex would be a system of movement existing relative to all other vortices acting as space. They would all be relative to the speed of light because that is the speed of motion underlying them all. This principle also resolves the dilemma of how a vortex can both extend in and create three dimensions. Each vortex extends in the three dimensions created by every other vortex in existence.

Energy is divided into bits and each is a particle of activity relative to all the other bits. Each part of the Universe depends upon every other part for its existence. The mutual interdependence of every particle of energy is the universal principle: *Love thy neighbour as thyself.*

At a quantum level every part acts for the whole and all particles of energy are totally interdependent. Every particle of energy exists relative to every other particle. In his special theory of relativity Albert Einstein was expressing the universal law of Love.

Just look at the key principle in Einstein's relativity that: *The observed velocity of light is independent of the velocity of the observer.*

An observer can only measure the speed of those photons of light, which reach him. As they do they would be travelling in his bubble of space. Because this moves with him wherever he goes, his measure of the speed of light would be independent of his own movement toward or away from the source of light.

To understand this, imagine you are in a car measuring the speed of a swarm of bees. You are driving alongside the bees until there is no relative movement between you and them and note your speedometer for their speed. It would be the same as yours. You then put your foot on the accelerator and race off ahead of them; turn round and head back toward the swarm. As you pass the bees in the opposite direction at the same speed, they appear to be going twice as fast as you are because the measured velocity of the bees would be relative to the velocity of your car. You note that bees hit your windscreen much harder when you drive against the swarm than when you drive with it.

Swarms of bees don't take kindly to being splattered into windscreens – even for a good cause like scientific research! Imagine a few angry bees find their way into the car. The velocity of the bees trapped in the car with you would be independent of the velocity of your car. No matter how hard you drive, in any direction, there is no way you could escape their stings.

In this analogy, the car represents your bubble of space, which moves with you. The bees represent the photons of light whose speed you are measuring. If the space, in which you measure the velocity of these flying particles, were moving with you, their measured velocity would be independent of your velocity: The vortex theory makes it very easy to understand Einstein's theories. Everything is simple and makes sense. Even time falls into place.

Time is an expression the relationship between things. Every movement or change in the Universe exists as a sequence of events relative to the other processes that are occurring around it. The interdependence of events in our world is what we experience as time. Every process takes time from the sequence of other repeating physical processes occurring around it. For example, we take our measure of time from the regular movements of the

Earth and Moon. The spin of the Earth on its axis gives us our days and its orbit round the sun gives us our years whilst the movement of the Moon gives lunar months.

This understanding of time enables us to comprehend Einstein's famous *Twin Paradox* theorem. If your time is the events going on around you, then should you accelerate the surrounding flow of events — your clock — would appear to slow down. If others were using you as their clock then, as a result of your acceleration, their time would appear to speed up.

Imagine you had a twin and you and your twin were the only things in existence. Your twin would be to you, space and time. She, in her turn would depend upon you for her space and time. Each of you would be space-time to the other. If you were to accelerate she, as your time, would have slowed down. From her point of view as her time, you would have sped up.

In Einstein's story the twin who underwent acceleration grew more slowly, in the dilated time. Returning from the journey the twin that accelerated was aged two whereas the sedentary twin was twenty!

Longevity caused by acceleration is borne out by the increased lifespan of particles undergoing acceleration in particle accelerators. At CERN, for example, when unstable muons (heavy electrons) were accelerated to nearly the speed of light, they lasted longer than muons at rest.

The flow of energy in one vortex, as a repeating sequence of events, would exist as a clock for the flow of energy in another vortex and vice versa. The acceleration toward the centre of the vortex would also affect time. If you were using a vortex as a clock then, with acceleration toward the centre, you would experience a dilation of time.

Imagine you were to suddenly shrink into the realms

of smaller space inside the atom. The minutes of your normal experience would first become hours, then days and then years. With acceleration toward the centre of the vortex seconds of time, measured by people of normal size, would come to be like years to you.

Imagine now you were to grow into a space-giant until planets were as footballs. As you leapt from star to star you would look down on the planet Earth and see it spinning like a top and whizzing around the sun. You might use it as a convenient timepiece taking terrestrial days as your seconds and human years as your minutes.

This vortex understanding of time reconciles the traditional dichotomy between religion and science over the time it took for life to come into existence.

According to the Bible, God created the world in seven days but science shows the evolution of the Universe and life forms occurring over billions of years. From the biblical perspective of a God looking down on the Universe, billions of human years would be as days. This conflict between religion and science is nothing more than a difference in point of view.

One could envisage Newton with his Bible imagining God looking on the Universe. Einstein, on the other hand, would take the position of the modern scientist looking up into the heavens. It is all a matter of relativity. Everyone sees the world from his or her own standpoint and judges others from that limited perspective. That is the primary source of conflict on Earth.

So many wars in history have been over different points of view. The only lasting resolution to conflict is the universal law of love embodied in Einstein's understanding of space and time; that everything exists and moves relative to everything else.

We all depend on each other. We are all interconnected. This is the law of time. To me, the parallels are clear between what Jesus taught on love and Einstein taught on time. The two men were describing in a different way, the mutual interdependence of everything.

Time is an expression of relative motion and change in the Universe. It is a consequence of the constant flow of energy through space. Time transforms everything because everything is fundamentally in a state of constant flux. Time will not allow anything to stay fixed forever because forever is but the cycle of coming and going; of birth and death, of dawn and dusk. The Hermetic principle, *that the rhythm of coming and going, growth and decay, life and death, expansion and contraction is universal,* is an expression of time. When Buddha said: *All are impermanent and doomed to change,* he was speaking of time.

If the vortex of energy sets up space-time there cannot be space or eternity existing outside the Universe. There can be no void existing as an eternal emptiness in which God dwelt for lonely aeons before creating the Universe.

If time is a consequence of the interactions of particles of energy, time cannot exist outside the Universe. There could not be a duration before the Universe existed or after its demise. There can be no beginning or end to the Universe because beginnings and ends invoke time which is a feature of the Universe of energy.

Because time is integral to the Universe an act of creating the Universe implies the existence of time before the beginning of time. The act of creating the Universe would be a sequence of events necessitating the prior existence of space and time. Space and time could not exist before the Universe because they are the Universe. Action outside the Universe is impossible because the Universe embodies all action that exists. No act is possible prior to the existence of energy because action is energy.

The creative act of God implies energy existed before it existed. Creation would not be some event in the distant past. Cosmic consciousness is creating, through active imagination, every particle in the Universe here and now.

There cannot be separation between the Universe and the source of the Universe because again separation is a property of space and time, which is a consequence of the existence of energy. To assume separation between God and the Universe or that God pre-existed the Universe would be to confer consequence on cause.

Debates on the origin of the Universe are as old as man himself. To quote Stephen Hawking:

The debate about whether and how the universe began has been going on throughout recorded history. Basically there were two schools of thought. Many early traditions, and the Jewish, Christian and Islamic religions, held that the Universe was created in the fairly recent past...On the other hand there were people such as the Greek philosopher Aristotle who did not like the idea that the Universe had a beginning. They felt that would imply divine intervention. They preferred to believe that the universe had existed and would exist forever...in 1781 the philosopher Immanuel Kant wrote a monumental and very obscure work, 'The Critique of Pure Reason', in which he concluded that there were equally valid arguments both for believing that the Universe had a beginning and for believing it did not. As the title suggests, his conclusions were based simply on reason...

I side with Aristotle, believing the Universe had no beginning and will have no end but for me that does not preclude the divine. Sir Fred Hoyle F.R.S., one of the greatest scientists of our age wrote a book called *The Intelligent Universe*. For me that title says it all. The eternal Universe is intelligent and the intelligence of the Universe is God.

Calder Nigel, *Einstein's Universe* BBC Publications, 1979

Clerk R.W. *Einstein: His Life and Times* Hodder & Stoughton, 1973

Hawking Stephen, *Black Holes and Baby Universes,* Bantam Books, 1993

Hoyle Fred, *The Intelligent Universe,* Michael Joseph, 1983

Melchizadek Drunvalo, *The Ancient Secret of the Flower of Life* Vols 1 & 2 1985-94

Richards, et al, *Modern University Physics,* Addison-Wesley 1973

Roland Paul: *Revelations: The Wisdom of the Ages,* Carlton 1995

Chapter 6

Intelligent Evolution

In him was life, and the life was the light of men.

John 1:4

Atheists and Christians have taken entrenched positions in their beliefs about the origin of life on Earth. Darwin's theory of evolution is the main platform of atheism today and Christians are fighting in the courts to have the Bible story of creation taught in schools.

According to the Bible, God created the Universe, the Earth and all living things upon it including us. This was achieved in a week. From the Bible, Christians have dated the epic week; it occurred about six thousand years ago.

Atheists believe that life on earth originated through an evolutionary process. For them happenchance led to life. They contend that atoms and molecules collided at random and after billions of years of haphazard interactions, became biochemicals. These biochemicals then formed a primordial soup and by fortuitous accident got strung out in the incredible DNA double helix. A long series of fortunate mistakes led to the genetic sequencing of the DNA and its replication in RNA. By jolly good fortune the RNA acted as a template for amino acids to form protein. These massive complex molecules then jostled about in the slime and 'hey presto' membranes and enzymes came into being, as the proteins fumbled by fluke into living organisms.

Despite the fact that DNA is remarkably stable, and that mistakes in copying are exceedingly rare and usually disadvantageous, biologists still insist that accidents in the replication of DNA, tested through natural selection, led to the remarkable diversity of life on Earth. The culmination of this process, dependant entirely upon luck, was the evolution of the human brain which endows us with the ability to think and be self conscious. Thus at the end of evolution did intelligence emerge. Not!!

Ralph Waldo Emerson said: *Shallow men believe in luck. Strong men believe in cause and effect."*

Christians believe the cause of life on Earth was God. Atheists believe there was no cause. I believe the cause of life is intelligent evolution.

Every argument begins with a fundamental assumption. Christians assume the existence of an all powerful, all knowing God who designs and creates everything. Atheists assume the there is no God, only matter and energy which gradually evolve into everything culminating in intelligent people like you and me.

My assumption is based on the fact that everything is energy. Modern physics has shown there is nothing underlying energy. The vortex of energy has revealed the illusion of materialism. This led me to conclude primary particles of matter are acts rather than material entities; that particles of energy are more thoughts than things. A mind is a body of thought. The Universe as a body of energy therefore appears to be a mind which places consciousness and intelligence as fundamental to reality. My premise is simple. If intelligence is the bedrock of reality it cannot be excluded from the evolutionary process.

Theists can satisfy their beliefs by assuming the Universe is the mind of God. Monotheists can define God as the universal consciousness imagining everything into

72

existence and my arguments for the indivisibility of consciousness define this God as one. Deists and pantheists can choose the conclusion that there is no external God but rather the Universe itself is conscious and intelligent; that God is the collective intelligence of the Universe; every quantum of energy being a thought form and every particle of matter a memory. Agnostics can continue to say they don't know and atheists can continue to say 'no!'

Ultimately it doesn't really matter what people believe so long as they are happy and tolerant toward other people's beliefs. My thesis for intelligent evolution is more belief than fact but if it makes sense to you, you can believe it too!

The assumption that the quantum is more a thought than a thing places mind at the beginning rather than the end of the equation for life. If particles of matter are more memories that material entities then intelligence could underlie molecular activity. Rather than life and mind being a consequence of chemistry, life could be viewed as the expression of the universal mind in chemistry and biochemistry. Through the non-materialistic, 'intelligent' approach to physics it is easy to conceive of intelligence underlying the vast edifice of evolution.

In *The Intelligent Universe*, Fred Hoyle reasons that evolution must be driven by intelligence. He said he preferred the Greek concept of the gods as managers in an already existing Universe to the Christian belief on a single God creating the Universe and everything in it. Hoyle considered it a vast unlikelihood that life could have evolved from non-living matter without the intervention of intelligence. To quote this brilliant scientist:

"...it is apparent that the origin of life is overwhelmingly a matter of arrangement, of ordering quite common atoms into very special structures and sequences. Whereas we learn in

physics that non-living processes tend to destroy order intelligent control is particularly effective at producing order out of chaos. You might even say that intelligence shows itself most effectively in arranging things, exactly what the origin of life requires."

If intelligence operates at a quantum level, then there could be consciousness underlying the atom, experiencing atomic interactions. Molecules may hold the memory of their combinations. Through trial and error intelligence could be evolving alongside biological systems. Intelligence also allows for information gained to be passed onto other systems.

The concept of 'systems intelligence' lies at the heart of Rupert Sheldrake's revolutionary idea of morphic resonance. But as he says, all evolutionary theories are speculative:

"Very little is known or can ever be known about the details of evolution in the past. Nor is evolution readily observable in the present…With such scanty direct evidence, and with so little possibility of experimental test, any interpretation of the mechanism of evolution is bound to be speculative."

If the Universe is a mind it would be learning from every event that occurs within it. Evolution could represent this learning process. If the Universe is a mind it would also be imaginative and creative; even artistic. I like to speculate that intelligent evolution is the mechanism of creation and many living organisms are pure art. Look at an exotic bird or a spectacular butterfly, a gorgeous flower or a colourful fish. In science we may search in vain to explain the beauty in the biosphere. In art no explanation is necessary.

Intelligent evolution makes sense as the 'creative' means through which universal intelligence develops the multitude of life forms that populate the world. An intelligent Universe could be evolving through life, learning and growing, breaking things down and then building them up again better than before. The extraordinary diversity and dazzling beauty we witness in life screams 'intelligence' but we don't have to accept that intelligence as the designer God portrayed in the Bible.

In 'The God Delusion' Richard Dawkins defines the 'God' hypothesis: *There exists a super human, supernatural intelligence who deliberately designed and created the universe and everything in it including us.*

In this delightful book he advocates an alternative view; *any creative intelligence of sufficient complexity to design anything, comes into existence only as the end product of an extended process of gradual evolution.*

Somewhere between these two contentions lies the view: *Creative intelligence evolves along with everything else in the Universe. There is no overall design just order emerging from chaos as intelligence learns through an evolutionary process.*

The idea of a creator God that is perfect at the beginning is pure speculation. Belief that God is only good doesn't fit with experience in the real world. Life can be cruel. A ruthless process of natural selection is closer to the truth. Darwin makes more sense to me than Genesis where everything is perfect from the start.

Through the Bible we are led to believe that humanity soiled the perfection. We made a terrible mistake which we can never correct; only God can redeem us.

I prefer to believe that through life the Universe is attaining perfection. My reply to theists and atheists is; *"Perfection lies at the end of evolution".*

Most people aim for perfection in their daily lives. Theorists like me put out ideas and see if they work. If they don't we adapt them until they do, or drop them and try something different. Evolving theories are subject to criticism so that only the best survive. But all the time we are in the process. The search for the perfect theory would never happen without the theorist. Consciousness and intelligence are never excluded from the human evolutionary process. Why should it be different for the 'Intelligent Universe'?

As we take rejection in our stride and learn from our mistakes we modify and make choices. There are lucky breaks but for the most part it is a learning process taken step by meticulous step. *As below so above, as above, so below;* why would it be different for the thoughtful Universe?

Authors like myself never find it easy to spot errors. We rely on editors and critics to point out the flaws in our work. That selection process can be painful. Natural selection is harsher. If you are not up to par you end up as lunch.

If I think hard for a solution to a problem I rarely find it. Solutions come more often in a creative flow where haphazard ideas are freer to come and go.

The idea of happenchance behind evolution, so dear to atheists, is a vital contribution to understanding the process. Random events are essential to 'creative evolution' because randomness allows freedom into the system. Creativity goes with freedom not repression.

Einstein said *"God doesn't play dice."*

If God is collective universal intelligence, it is through dice that God plays, as chance is an essential ingredient in intelligent evolution.

The Nobel Prize laureate Jacques Monod stated: *"Chance alone is the source of every innovation, of all creation...Pure chance, absolutely free but blind, at the very root of the stupendous edifice of evolution: this central concept of modern biology is no longer one among other conceivable hypotheses. It is today the sole conceivable hypothesis."*

Richard Dawkins argues the same viewpoint. In the *Blind Watchmaker* he contends there is no overall sense of direction in life and certainly no designer. The process of natural selection which drives evolution is completely blind in his opinion. Dawkins accepts that the origin and evolution of life does require a whole sequence of highly improbable events. But he believes over millions upon millions of years, 'near miracles' could have occurred. Evolution is cumulative. Each tiny step would build on all the others.

Dawkin's insightful concept of cumulative steps in the process is a major contribution to understanding evolution. Randomness has a vital part to play in the natural selection process for testing new biological systems. However, the process may not be blind.

The Western mind fails to see the underlying order in apparent chaos and perfection in the operation of chance. Not so the Chinese mind.

To quote Carl Jung from a forward to the 'I Ching': *"The Chinese mind, as I see it at work in the I Ching, seems to be exclusively preoccupied with the chance aspect of events. What we call coincidence seems to be the chief concern of this peculiar mind and what we worship as causality passes almost unnoticed."*

For intelligent systems random events and probabilities create opportunities for creativity and originality. The actions of quantum intelligence may appear chaotic but conscious evolution allows them to be watched. Under the trauma of natural selection they are tested and intelligence allows for improvements to be made. Consciousness also allows for cooperation to enhance the chance of survival in harsh circumstances. Bacteria converging to form organelles in a rudimentary cell could be an intelligent move rather than a lucky break.

Intelligence enables us to understand how the DNA double helix emerges from a precise, systematic arrangement of nucleic acids. The faithful replication of DNA genetic codes into proteins is communication. What communication occurs without consciousness?

Imagine bacteria consciously contributing their genetic strands as they congregate and transform into the organelles of primordial cells. Cells may be conscious as they cooperate in the tissues of a body.

Each rung on the ladder of intelligent evolution could represent a step by step increase in order of intelligence, each representing a more highly evolved collective of the quantum particles of intelligence. Maybe the human is not the only conscious and intelligent being? Maybe every cell in the body is conscious and intelligent, only at a different level. Consciousness and intelligence have been shown to exist in all living organisms. There is evidence that plants and animals respond to humans. People who live with animals are often well aware of animal emotions and their response to human emotions. As detailed in the 'The Secret Life of Plants' the great physicist Bose experimented with plant perception and technology exists to amplify micro-electrical signals in plants that will register plant response to human thoughts emotions.

With conscious awareness existing in every atom and molecule, organelle and cell, nowhere would consciousness be missing. Everywhere intelligence would be operating because the quantum of energy is a thought form, a particle of the universal mind. If atoms and molecules are memory forms within the Universe, macromolecules in living organisms would be stores of the cumulative effect of intelligent evolution. We witness the arrangement of cells in multi-cellular organisms and note that in these increasingly complex structures ever more information can be stored, expressed and assimilated. Life forms are the living memory of the Universe.

Energy is not uniformly distributed in the Universe. It is concentrated in galaxies and clusters of galaxies, in stars and planets and on the planets in complex life forms. The concentration of particles of energy into increasingly complex bodies is a concentration and organisation of quantum particles of intelligence which allow for ever more powerful expressions of reason. Humanity is the culmination of this process on Earth.

And the Word became flesh…

John 1:14

Bird Christopher & Tompkins, *The Secret Life of Plants*, Allen Lane 1973

Darwin Charles, *Origin of the Species*, Penguin

Dawkins Richard, *The Blind Watchmaker*, Penguin, 1988

Dawkins Richard, *The God Delusion*, Bantam Press, 2006

Hoyle Fred, *The Intelligent Universe*, Michael Joseph, 1983

Monad Jacques, *Chance and Necessity*, Fontana, 1972

Sheldrake Rupert, *A New Science of Life*, Paladin Books, 1987

The Holy Bible: New King James Version, *Nelson Bibles*

Wilhelm Richard, *I Ching*, Routledge & Kegan Paul, 1951

Chapter 7

Understanding Infinity

"Let us hope that out of this chaotic riddle will come a profound and simplifying answer. We may be likened to those who knew only Ptolemy's complex description of the solar system. What we need is a Copernicus to assimilate and interpret the data with a generalisation which will not only solve the riddle, but lift our sights to levels we cannot now foresee."

Richards et al Modern University Physics

T he question with the vortex is: Where is all the energy coming from? Where is it going to? The vortex can only exist if the concentric spheres of energy are continually growing or shrinking through it. The vortex cannot be an infinite source or sink of energy. Energy is neither created nor destroyed. The vortex must be a part of a cycle of energy.

In nature, things flow endlessly when they are part of a cycle. The water cycle is an example. A whirlpool can exist indefinitely in a river so long as water is continually flowing through it. An endless flow of water in the river is possible when rains feed the springs. The river is part of the water cycle with innumerable raindrops at one end and a single ocean at the other. The water cycle is divided into two halves. One half is the flow of water from the springs to the ocean through the rivers. The other half is the flow from the ocean to the springs via the rain.

The unending flow of energy in the vortex could be part of a universal cycle of vortex energy similar to the water cycle. The innumerable centres of protons would be like raindrops. There could be a single largest sphere of space that can be likened to the sea. Just as the oceans collect all water in the world so the largest sphere of space would contain the vortex energy extending from every particle of matter in existence

The flow of energy through the vortex could be likened to streams and rivers carrying the energy from where it rises in the springs to its destination in the ocean. The evaporation of water from the ocean to form clouds and then rain to maintain the springs would represent the other side of the vortex cycle of energy - the flow of energy from the maximum sphere of space back to the centres of the protons again.

Because we are formed out of the vortices of energy in matter we can only be aware of one half of the cycle of vortex energy. We are like fish in a river.

Fish, immersed in the waters of the oceans and rivers know only oceans and rivers. They have no knowledge of clouds and rain. As raindrops fall into the surface of a river, a super intelligent fish could become aware of them as another form of water apart from the river. The smart fish could speculate that the rain drops belong to the missing half of the cycle predicted to explain the endless flow of water in his river.

The discovery of particles of antimatter, by Carl Anderson of the California Institute of Technology, was the equivalent of a clever fish discovering rain. Antimatter appears in the world of matter like drops of rain penetrating the surface of the river. From antimatter we can speculate the existence of another world through which vortex energy flows to form an endless cycle.

The observed world of matter may be but half of the Universe. There may be another half formed out of anti-matter. Between these two halves, connected in a circle, space would be infinite.

Each half of the Universe is represented by a loop of the infinity sign. Through these two halves of the Universe, vortex energy could flow in an endless cycle.

The vortex of energy could be pictured as a funnel. The centre of the vortex is the smallest point of space.

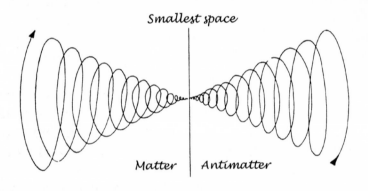

Smallest space

Matter | Antimatter

This is represented by the small end of the funnel. Energy would spiral in to form the funnel representing a negative charged particle of matter. Then, passing through the minimum-space point as though it was a tunnel, the energy would spiral out to form an identical vortex funnel. This would represent a positive charged particle of anti-matter.

This picture suggests that every vortex in the world of matter would share a common centre with an identical vortex in the world of anti-matter. The anti-matter vortex would have the same mass as the vortex of matter but it would be flowing in the opposite direction. The two vortices would be connected like Siamese twins. Every particle of matter in our world would share a common smallest-space point with a particle in the world of anti-matter.

The idea that there is an anti-matter half to the physical Universe is supported by the 'mirror-symmetry of matter and anti-matter'. Practically every particle of matter has a twin, a mirror-symmetrical anti-particle. If every particle of matter is connected to a particle of anti-matter then every form and action in our world must be mirrored in the world of anti-matter. Actions could only be mirrored if every photon of light in our world were shadowed by a photon in the world of anti-matter. For every photon in the world of matter there must be a 'shadow photon' in the world of anti-matter.

If this idea is right at this moment there is a mirror image of you in an anti-matter world reading an anti-matter version of this book. The question is, between you and your anti-matter double, which is the real you and which is a mere reflection?

To complete the cycle of vortex energy, the worlds of matter and anti-matter must share the same largest sphere of space through which vortex energy from both worlds would continually flow. In the vortex cosmology, the infinity of space is a circle in which the worlds of matter and anti-matter are connected in the smallest and largest realms of space. This is the dimension of infinity.

The dimension of infinity is a dimension of size; a dimension of shrinking or growing. You traveled in this dimension as you grew from a single cell into a full grown adult. I call this dimension, the *Alician Dimension* after Alice of *Alice in Wonderland* who traveled in it when she ate the *eat-me-biscuit* and drank the *drink-me-drink*. When Alice shrank in size she entered a looking glass world where she met the twins who agreed to do battle... So it

was, with extraordinary insight Lewis Carroll anticipated the mirror symmetrical world, reached by change
in size, where twin particles of matter and antimatter
have a built in agreement to do battle to annihilation if
ever they chance to meet.

To understand the matter-antimatter structure of the
Universe, take an imagi-
nary journey in the Alician
Dimension. To begin visu-
alise the centre of a proton
in your head expanding as
a minute bubble. The
sphere of energy grows to
make way for another
bubble expanding out of
the proton centre. New
bubbles of energy appear
from the point of singularity to create the steady
stream of concentric
spheres that form the proton.

In your imagination join one of the bubbles and travel
with it through the Universe. The energy in your ever expanding bubble becomes imperceptible beyond the
atomic nucleus but the spheres of energy from the other
particles in the nucleus merge with it to form a sphere
which you can perceive.

This sphere of energy grows out from the nucleus as
the original sphere grew out of the proton. There are
larger spheres ahead of you and smaller ones behind, all
expanding on the same journey. As your bubble of energy grows out into the atom, you see concentric spheres

of energy traveling in the opposite direction. They are shrinking into the electrons orbiting the nucleus.

You are now surrounded by millions of atoms in a DNA molecule then swelling past mitochondria factories, busy membranes and engorged vacuoles, you find yourself outside a brain cell. Still growing, the sphere of energy is first as big as your brain and then your body. It is now no longer a sphere but is an extension of your shape and contains vortex energy from every subatomic particle in your body.

Your vortex energy merges with the space bubbles of other people as it continues to grow. They shrink then vanish as your bubble of energy grows to enclose the entire Earth. The sphere is 13 thousand kilometres in diameter and contains vortex energy from every particle in the planet. The Earth is vast beneath you, but then as your sphere of energy expands, the globe shrinks to a blue and white marble suspended in space then is gone. Your sphere of vortex energy is now 12 million kilometres in diameter.

Your orb surrounds the entire solar system. Jupiter burns red and the sun, bright gold. At 100 trillion kilometres in diameter, your bubble of space includes the three stars of Alpha Centuri, Sirius and Barnard's single star. Still growing, at a million, trillion kilometres your bubble has become a mighty spheroid containing the vortex energy extending from the entire Milky Way galaxy. Soon your bubble of energy has grown to encompass a cluster of some twenty galaxies including the Milky Way, the Magellanic clouds and Andromeda, bursting with billions more stars.

The cluster shrinks from sight as your bubble grows to enclose hundreds of galaxy clusters including Virgo with thousands of galaxies and Canes Venatici with thousands more. Expanding out of this super-cluster, when your orb is a 100 billion, trillion, kilometres it envelopes the super-clusters, Peruses, Hercules and Indus. At 200 billion, trillion kilometres you encompass energy from thousands of super-clusters. At 300 billion, trillion kilometres you begin to encounter quasars, radiating two hundred times the energy of a star. At 400 billion, trillion kilometres you have swept up countless millions of quasars and with a diameter of 500 billion, trillion kilometres you are the largest sphere of space containing vortex energy extending from all matter. This is the outermost frontier of the Universe; the largest sphere of space. Here your sphere stops expanding and begins to shrink.

Instead of travelling with the drifting galaxies you are now moving against them. Faint specks of light grow into mighty super-clusters Peruses, Hercules and Indus that grow up and expand past. You seem to shrink back the way you came as Virgo appears and expands into a thousand galaxies then as Andromeda flies past you are enveloped once again by the Milky Way.

A single point of light grows toward you. It is the Sun. Jupiter, Mars and Saturn then swell past as you shrink towards the white and blue marbled Earth expanding to meet you. Viewing continents amidst the clouds, you are drawn to your own country. A familiar town appears as you contract toward a human body. At first it is minute but as your sphere shrinks it grows bigger than a giant. Hair stands like trees as you shrink into a single brain cell then into its nucleus. A coil of DNA grows up as you vanish into the double helix strand and then into an atom. Spheres of energy are expanding out of its orbiting vortices which are positrons, not electrons!

You have just passed through a world of antimatter and, drawing ever closer to the antimatter atomic nucleus you shrink into an anti-proton and again becoming a single sphere of energy. Compressed into the single point of its centre, you pass through the singularity and begin to grow again. As the sphere of vortex energy expands out to form a proton, you are back where you started.

The cycle between matter and antimatter is the infinite cycle of space. This understanding of infinity leads to an account for gravity and the accelerating expansion of the Universe.

Richards, et al, *Modern University Physics* 1973 Addison-Wesley.

Chapter 8

Gravity and Anti-gravity

I am hopeful we will find a consistent model that describes everything in the Universe. If we do that, it will be a real triumph for the human race.

Stephen Hawking

Matter is attracted to antimatter. Your body experiences a pull through smallest-space, towards antimatter in the other world. The pull is not between you and your double because you belong to the same system of energy. The pull would occur between you as a body of matter and all the antimatter in the mirror image Earth.

Acting through the centres of all the vortices, this pull is centralizing force that causes particles to conglomerate in both worlds. Because energy accelerates into the vortex we experience this centralizing pull on matter as acceleration. This pull is gravity.

In the vortex cosmology there is a polar opposite to gravity. This is a pull on matter from the world of antimatter, through the maximum sphere of space. Acting through the largest sphere of space, it affects the largest bodies of matter. It would cause the decentralisation of matter because it comes from the outermost reaches of the Universe. The pull would therefore cause galaxies to fly apart from one another — which appear as the expansion of the Universe. The electric pull between matter and antimatter causes them to accelerate together therefore the expansion of the Universe would be accelerating.

At present, it is widely believed that the Universe is expanding because of a push on matter, caused by a big bang at the beginning of time. In the vortex cosmology the expansion is caused by a pull rather than a push. A pull is the antithesis of a push so the vortex account for universal expansion is an antithesis to the big bang theory.

The current phase of the Universe may have begun with a big bang. However, on its own it is an unsatisfactory account for the expansion of the Universe. The big bang theory predicts that the expansion of the Universe is either uniform or slowing down. Prior to the 1990's it was assumed that the Universe was expanding at a uniform rate. However, supernova explosions in distant galaxies reveal the expansion of the Universe isn't slowing down but is speeding up.

To quote the Astronomer Royal, Sir Martin Rees: *"An acceleration in the cosmic expansion implies something remarkable and unexpected about space itself: there must be an extra force that causes a 'cosmic repulsion' even in a vacuum. This force would be indiscernible in the Solar System; nor would it have any effect within our galaxy; but it would overwhelm gravity in the still more rarefied environment of intergalactic space. Despite the gravitational pull of the dark matter (which acting alone would cause a gradual deceleration), the expansion could then actually be speeding up."*

A major weakness in current astronomy is the account for the apparent absence of antimatter. Lord Rutherford and Frederick Soddy's law of Conservation of Charge requires that antimatter must have been created in exactly equal proportions to matter in the birth of the physical Universe. To account for the apparent absence of antimatter in the Universe scientists speculate that there must have been slightly more matter than antimatter to begin with. Most of the matter annihilated with all of the antimatter, so the residual matter is what we perceive as the Universe today.

In the vortex cosmology an account is provided for antimatter. It exists now in exactly equal proportions to matter. This account is essential to the vortex theory to complete the cycle of energy through vortices in matter.

Prediction of the antimatter half of the Universe leads to an explanation for gravity. It also explains why distant galaxies are accelerating toward the outermost reaches of space; why the further galaxies are from us, the faster they are moving away from us.

If a galaxy of matter were to accelerate toward a galaxy of antimatter and vice versa, eventually they would meet and annihilate. There is evidence of annihilation going on at the outermost reaches of the Universe. The furthermost things from us are quasars which radiate about two hundred times the energy of normal stars.

Since stars produce energy through nuclear fusion, which consumes just over 0.7% of their total proton mass, a process that produces approximately 200 times this radiation of energy would consume 100% of the mass of a proton. The annihilation of matter and antimatter produces this amount of energy.

Matter and antimatter meet in the smallest as well as the largest regions of space. Annihilation should occur in the regions of greatest centralisation of space as well as in the regions where space is most decentralised. In fact, quasars occur in the central regions of many galaxies —regions where matter would be expected to be most centralised.

Quasars would form in a galactic centre as dead and dying stars converge due to gravity. As stars exhaust their nuclear fuel and cool down, they eventually fall in on themselves due to gravity. First they form a red giant then collapse into a neutron star or a white dwarf star. In the demise of very massive stars, gravity can be sufficiently strong for a black hole to form. A massive black hole in a galactic centre could pull active stars as well as neutron and white dwarf stars with its gargantuan gravity.

The black hole represents the greatest density of matter possible and as it forms, an antimatter black hole would form beyond its centre, through the smallest realms of space. As particles of matter converge into smallest space they would meet particles of antimatter and annihilation would occur. To begin with, the energy released by this process could not radiate from the black hole because it would be acting as a colossal 'proton like' vortex, capturing the energy of annihilation in its spiral space-path.

However, with annihilation, the mass of the black hole would diminish and its gravity would drop. The ability of the black hole to capture energy would decrease. With progressive annihilation the energy contained by the black hole would increase. Eventually there would have to be a crossover point between the diminishing gravity and the increasing energy of annihilation. Energy would then escape and radiate away. At that point the formation would become a quasar. A black hole would be only a temporary astronomical event, occurring as a step in the transition of stars into quasars.

In the cosmology of the vortex the black hole appears, not as an astral grave, but as a stellar pupa, holding the metamorphosis of the 'caterpillar' star into the Quasar 'butterfly'.

Stars are non sustainable. Eventually every star in the Universe will become exhausted and black holes will increase in galaxies to swallow the dead and dying stars. Accelerating galaxies will encounter their antimatter doubles and annihilate each other out of existence.

In the inexorable pull between matter and antimatter through the smallest and largest realms of space — through gravity and universal expansion — eventually every vortex of matter and antimatter in existence will be destroyed. Everything will return to light. However, with destruction of vortices there would be a collapse of space in which the light moves. Energy would be forced to spin into the collapsing space and in spinning would be transformed back into mass.

Eventually all the energy of the Universe would be compressed into a point of singularity; a moment of absolute darkness, the night of the Universe. The newly formed matter and antimatter would be crushed with the light into an almighty bomb. With a big bang, a new Universe would be born. A new day would break with a brilliant light as energy and particles explode into new space and a brand new time. There is no beginning or end to the Universe only death and rebirth in the night and day of Brahma.

In the night of Brahman, nature is inert and cannot dance till Shiva wills it: He rises from his rapture and dancing sends through inert matter pulsing waves of awakening sound and lo! Matter also dances, appearing as a glory around Him. Dancing, He sustains its manifold phenomena. In the fullness of time, still dancing, he destroys all forms and names by fire and gives new rest. This is poetry but none the less science.

Ananda Coomaraswamy

Ash David, *The New Science of the Spirit,* College of Psychic Studies 1995

A. K. Coomaraswamy, *The Dance of Shiva,* Noonday Press 1969

Richards, et al, *Modern University Physics* 1973 Addison-Wesley.

Rees Martin *Just Six Numbers* Weidenfield & Nicolson 1999

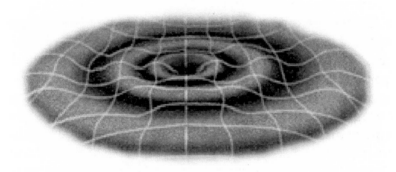

Chapter 9

The Love of Shiva and Shakti

The principles of masculine and feminine are found at every level in the Universe.

Hermes Trismegistus

P articles of energy are imagined into existence with predetermined forms. The lines of movement exist in the form of vortices or waves. The form is a persistent memory. If change of form is imposed on a particle it will maintain the change only whilst the force of change remains. When the force of change is removed it will revert immediately to its original 'natural' form. These are the axioms of the new physics of consciousness which lead to an entirely new approach to quantum theory; an approach based on love, the love of Shiva and Shakti.

The axioms are not drawn from experiment, they are freely invented because, as Einstein said: *"The axioms of a scientific theory cannot be drawn from experiment, they must be freely invented."*

The wave and vortex are perpetual forms of energy. If the motion in the vortex and wave particles were not perpetual the Universe would have vanished a long while ago, so this fact is self-evident. The stability of the form of motion in each particle of energy is also self-evident. If this were not so then the Universe would be completely amorphous.

One example of the stability of the vortex is the esti-

94

mated 10^{33} year life span of protons. The durability of the waveform of energy is also evident from the ability of photons to survive journeys through space lasting billions of years. The kinetic inertia of the undulating dance of Shiva and the static inertia of the swirl of Shakti come from the innate stability of waves and vortices of energy.

I have proposed two laws to express the innate stability of the two fundamental forms of energy and govern their interactions. These are *The Quantum Laws of Motion*. The quantum laws of motion are an application of Newton's laws of motion at the quantum level.

If everything were formed out of particles of perpetual motion then the most fundamental laws of the Universe have to be laws of motion. Also, if particles of movement are inherently stable then stability should be a primary feature of these laws.

Newton stated, as his first law of motion: *Every body will continue in its state of rest or uniform motion in a straight line unless compelled by some external force to act otherwise.* Children illustrate this fundamental law of inertia. It usually takes force to stop them doing what they are doing and get them to do something different. Orders for change, such as, *"stop playing it is time for bed,"* almost invariably meet with resistance. This is embodied in Newton's second law: *Action meets with reaction.*

Newton based his laws on the behaviour of large bodies of matter. They have to be modified if they are to apply at a quantum level. For example, energy is not motion in a straight line. The particles of energy, described as quanta or photons, possess wave motion. No external force has ever been observed to cause their motion to deviate from a straight into a wavy line. Wave-motion is the fundamental inertia of these 'radiant particles' of energy, which can be described as wave-kinetic inertia.

95

A vortex particle of matter will maintain its relative state of rest unless another particle acts upon it to propel it into motion. This has been described already in terms of static inertia and mass. Vortex motion is the fundamental assumption of the theory so it is not necessary to explain 'why vortex motion'. It is no more possible to account for why the line would occur as a spiral than why it would exist in wave form.

The wave and vortex forms of energy are perpetual and fundamental to the universe. This axiom is expressed in the first quantum law of motion.

I. *The First Quantum Law of Motion:* Particles of energy maintain their state or form of motion in spin or wave unless a change is forced upon them.

The innate stability of spin and wave lead to the second law.

II. *The Second Quantum Law of Motion:* When the force of change is removed, the particle of movement will revert immediately to its original form.

The two quantum laws of motion are illustrated by a coiled spring. The form of a spring is altered by the application of a force but on release of the force, it reverts to its original shape.

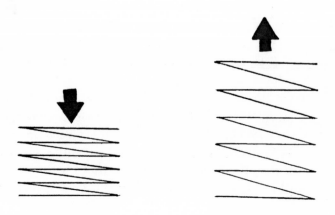

Springs have a memory of their shape. They can be distorted but revert to their shape memory when the cause of distortion removed.

The quantum laws of motion explain Cecil Powell's mesons. As kinetic energy in the motion of the cosmic ray particle drives through the vortex particles in the nucleus of the silver atom it is transformed into short lived vortices identified by Powell as mesons. This is the first law in action.

The decay of the mesons, within billionths of a second, into radiant energy is the second law in action. On exiting the nucleus as there is nothing to maintain the meson energy in vortex motion they unravel and reverting to wave-kinetic energy they radiate away as photons.

This production and decay of mesons can be appreciated by means of a doughnut analogy. The doughnut mould represents vortices in the nucleus of an atom. Kinetic energy is represented by the batter. When batter is forced through a doughnut mould it takes on the shape of a doughnut. So it is when kinetic energy is forced through a subatomic proton vortex it takes on the form of the vortex, which appears as a meson. This illustrates the first quantum law of motion.

If the batter in doughnut shape falls onto the floor it will immediately revert to a mess of batter. When the meson swirl exits the proton it immediately unravels, reverting to the wave form and radiates away — in less than a billionth of a second!

In physics we are taught that motion results from the application of a force. This is putting the cart before the horse. All forces are derived from motion; the motion we call energy. Motion is the prime reality and forces result from the form of motion.

A wave-particle of energy will force a vortex into

wave motion. A vortex-particle of energy can force a wave train of energy to be transformed into mass. These interactions between vortices and waves are described in a branch of the new physics of consciousness called *Capture Theory*.

When the wave dance of Shiva is captured by the spin of Shakti a vortex particle of matter captures a wave particle of light. This energy capture underlies kinetic energy and transformation of energy into mass and mass into energy.

Energy capture occurs according to the quantum laws of motion. You could imagine the quantum laws defining the steps of the dance between Shakti and Shiva. The result of energy capture is that Shiva and Shakti become entwined. Their love play is fundamental to all processes in the Universe.

The operation of capture is simple. Shakti takes no action. She is simply irresistibly attractive to Shiva. Shiva takes the action of dancing into Shakti. Fertilization and gestation follows penetration of the female by the male. This fundamental law of the Universe applies right down to the level of the quantum.

The vortex presents itself to a wave train of energy as a spiral 'space path'. The radiant energy follows the contours of space much as a car follows the bends in a road. Just as a car is forced to follow a curved path when it enters a round-about on the road, so the 'travelling wave-kinetic energy, following the contour of spiral space,

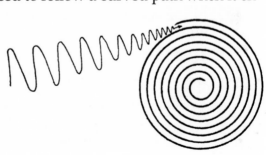

would move in toward the centre of the vortex.

As it travels in an ever-tighter spiral space-path, the wave train of energy would become captured. Closing in on the centre of the vortex the kinetic energy would be transformed into mass.

Vortex particles such as protons are sufficiently massive to completely capture entire quantum particles of wave-kinetic energy transforming them entirely into mass.

Some vortex particles — electrons — are insufficiently massive to completely capture an entire quantum particle of wave-kinetic energy. The electron has very low mass and therefore little static inertia. The light electron vortex would be propelled forward by the impact of a wave train of energy. Because the wave train is driving the electron forward, the wave train would be unable to drive completely into it.

The tip of the energy wave train would be tangled in the electron vortex and transformed into mass. By means of this transformation the electron would partially capture the wave-kinetic particle of energy.

The wave train of energy, which is not transformed into mass, could be imagined as sticking out of the vortex like a tail. This wave-kinetic energy would drive the vortex forward in wave motion. Through this interaction, the wave train and vortex would become a couplet of the two forms of energy. Each would impose its inertia onto the other. The wave train would cause the electron to be a wave-particle, as it is driven forward in wave motion and the vortex would cause the captured energy to add to its mass. This explains the wave-particle duality of electrons. Electrons are particles that move according to the laws of wave propagation.

The wave-vortex couplet could be imagined as a tadpole with the vortex as the head and the partially captured wave train of energy as the tail. Just as a tadpole swims in wave motion because it is propelled by the wave motion of its tail, because of the motion of its captured wave-kinetic energy, the vortex would move as a wave-particle.

The increase in mass of particles with acceleration can be likened to cannibal tadpoles growing strong as they consume their siblings and then eating even more of their unfortunate brethren because they are bigger and stronger. As the vortex particle absorbs more wave-kinetic energy to accelerate it becomes more massive and so requires even more energy to go through another increment of acceleration. Approaching the speed of light the increase in mass is faster than the acceleration.

According to the first quantum law of motion, whilst the wave-particle state persists, the electron would have mass slightly greater than its theoretical rest mass and it would also possess perpetual wave-kinetic energy. In the atom this is the ground state electron, which can orbit the nucleus indefinitely.

The ground-state electron can be illustrated by the image of a stable marriage. Just as in a marriage, each partner imposes their personality on the other, so it is that in the entwined Shakti and Shiva, the couplet of wave and vortex, each type of energy imposes its form of motion, or inertia, on the other; the wave form takes on mass and the massive form takes on wave propagation.

If the electron vortex absorbs another quantum of energy into her spiral space path she will move into an excited state and take a quantum leap into a higher orbit.

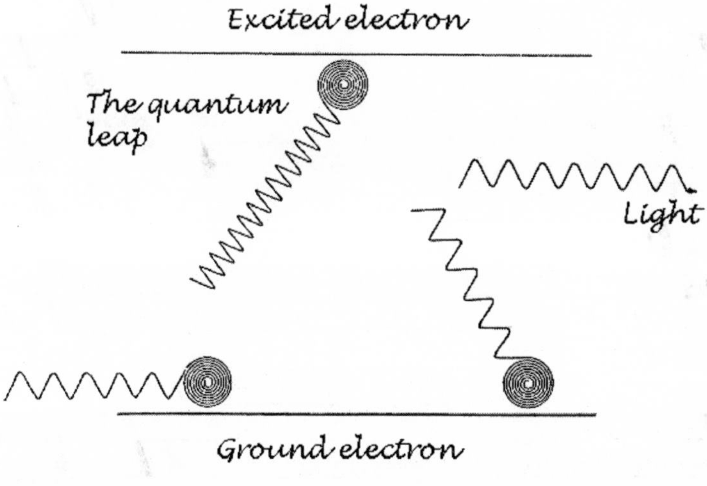

Excited electron

The quantum
leap

Light

Ground electron

The excited state electron is the equivalent of a wife taking a lover. A wife who gets the hots becomes excited. An electron that is heated becomes excited..

The triangle of husband, excited wife and lover is unstable. So it is the excited state of the electron is unstable. The lover may leave the wife. In that case she would be grounded with her husband. The electron may lose the energy of excitement as a photon of light. In that case it drops back from the excited state to the ground state. This happens all the time in a flame. The light emitted has a colour representing the frequency difference between the excited state and the ground state. In carbon atoms the energy difference between the excited state and the ground state is the frequency of yellow light. Candles emit yellow light because of particles of carbon in the flame.

Wives are fussy about their lovers. He has to have what it takes to lift her into an excited state. So it is a quantum has to have what it takes in frequency to lift the electron from the ground state to the excited state.

If the chemistry is right the excited wife may leave hubby behind and skip off into a new home with her lover. Electrons do that too. In the excited state electrons are prone to leave one atom for another. In chemistry this causes a chemical reaction and the basis of the chemical bond.

Some wayward wives nip back and forth between the old home and the new. Electrons do that too. Nipping back and forth between the old atom and the new they form a covalent bond between the two.

A willful wife may never return to the old home. Charge is all that's left between her and hubby. Electrons sometimes leave an atom for good and go off into a new one and all that's left between them is charge. In chemistry that is the basis of the ionic bond.

Hermes stated that ...*the principles of masculine and feminine are found at every level in the Universe...* If he were alive in the 21st century imagine how he would have been excited by the microcosmology of quantum theory. He would have seen in the wave and vortex the duality of male and female operating at the sub-atomic level. He might have visualised wave particles of energy as sperm. Then he would have proclaimed heat and light as the masculine form of energy. He may have depicted the vortex particle of energy as an ovum. It is spherical, static and receptive. Then he would exclaim that vortex particles in matter are the feminine form of energy.

In the Native American tradition, light is seen as masculine and matter as feminine — which is why Native Americans speak of 'Grandfather Sun' and 'Grandmother' Earth. The word 'matter' is derived from the Latin word *Mater* for 'mother'.

The trinity of father, mother and child occurs at every level in the Universe. In the *microcosm*, the waveform of energy is the primordial masculine and the vortex is the primordial feminine. The wave-vortex wave particle is the primordial child. In the trinity of science the masculine is represented in physics, the feminine in chemistry and the child in biology.

In the *macrocosm* Hermes could have seen the masculine represented in the star, embodying the chaotic fire of nuclear physics. This embodies Shiva manifesting the fire energy of destruction, glorious and terrible he will consume everything that comes too close and yet without Shiva there is no life because without kinetic energy matter is dark and inert. The explosive energy of Shiva represents the primordial rage of creation and destruction. The wave-energy of Shiva represents the nuclear fire of the stars. Shiva is expressed in the stellar plasma driving protons, electrons and neutrons with high-energy in the nuclear furnace of the sun. Within the stellar crucible of creation, in the destruction of stars, Shiva the destroyer works his alchemy in magnificent supernova explosions. There atomic nuclei are born but the atom has no form. The 'rudra' or rage of Shiva is chaos as his blazing dance lifts matter from the insatiable pull of gravity, the vortex grip of *Mater* the mother.

In the *macrocosm* Hermes might have seen the feminine represented in the planet. The dark and receptive planet embodies Shakti. The creativity of the Divine Mother is expressed in the atom. Within the cool and the dark of the planet, electrons settle into orbits around atomic nuclei. Far from the furious fire of the father atoms form and are nurtured by the mother. From her chemical interactions the subtle and majestic forms of creation unfold. This is chemistry. Shakti is Mother Nature embodying the creation of matter. She is author of life.

Shiva makes love with Shakti. She responds to the heat of His fiery embrace. In response to His seminal waves of light, her matter dances. On the planetary surface the sperm-like waves of light and heat fertilize the ovum like vortices in matter. Where matter meets light, where wave energy drives into vortex energy, chemistry gives way to biochemistry and biology, the child in the trinity of science, emerges. The word, the vibration, becomes flesh and a son is born.

Without Shakti, Shiva is lost in the heat of his rage. Without the balance of the feminine the masculine is destructive. Consider a man, roaming in the wilderness lost and dissipated, leaving his woman at home frustrated and fruitless. When men leave their women they go to war. Without matter, light radiates into space and is dissipated. Without Shiva, Shakti is frustrated. Without the heat of her beloved Shiva, his chaotic dance to channel, Shakti is cold and fruitless.

Without kinetic energy, suns die, planets drop to temperatures approaching absolute zero and particles of matter collapse into black holes. Nothing worthwhile happens when the wave and the vortex, the masculine and feminine forms of energy are separate. This is entropy.

The opposite of entropy is work. Work describes the dance of Shiva harnessed by Shakti into useful processes and forms. The marriage of Shiva and Shakti makes possible the interactions between matter, heat and light. In physics this is the science of Kinetics. Wave-vortex couplets enable the vortex to do worthwhile work and form useful structures out of atoms. This is evident in human society. When men marry women their masculine energy is channelled into work. When women harness men into work swords are fashioned into plough shears.

Chapter 10

Shiva's Fire

The dancing Shiva, depicted in bronze sculpture ...is surrounded by tongues of flame symbolising the destruction that brings about the end

A. K. Coomaraswamy

Shiva's fire is nuclear energy. This is the fire of stars that flood planets with heat and light. However, in the hands of man Shiva's fire has the potential to destroy civilisation. Understanding nuclear energy will not lessen the threat it poses humanity but it increases the credibility of the new vortex physics and thereby may help increase consciousness.

Surprisingly there is no satisfactory account for nuclear energy in physics. To quote Richard Feynman: *"Nuclear energy... we have the formulas for that, but we do not have the fundamental laws. We know that it is not electrical, not gravitational and not purely chemical, but we do not know what it is."*

The spiral structure of space within the massive vortex particles in the nucleus of the atom — the proton or neutron — would make it easy for energy to enter but practically impossible for it to leave again. Completely captured in the proton or neutron vortex, energy would then swirl around inside indefinitely.

Imagine the proton vortex as a crab pot and wave energy to be a crab. The crab can enter the pot but then it can't escape. A crab pot takes no action to capture crabs. In like manner proton vortex would not act to capture energy. The pot is passive. It is the crab that crawls into it. In the same way, the proton vortex would be passive and the energy would fly into it and so become captured. The crab trap captures crabs through the way it is structured. In like manner a proton would capture energy as its structure presents a spiral space-path to energy. A trap can continue to capture creatures until it is full. In the same way, a proton or neutron vortex could continue to capture energy until it is saturated with captured energy.

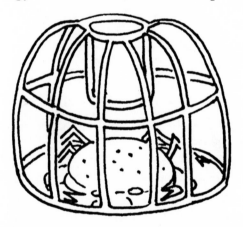

Protons and neutrons (collectively called nucleons) saturated with captured energy could also be likened to buckets full of water. If one bucket were placed inside the other the capacity to contain water would be reduced and so some of the water would spill out. If two nuclear particles were to converge, their capacity to contain captured energy would be reduced and some of it would be displaced.

The displacement of captured energy, due to the convergence of nucleons, accounts for nuclear energy. Because captured energy constitutes part of the mass of the proton, the loss of captured energy, due to the convergence of two protons — or neutrons — would be accompanied by a small loss in the mass of the protons. This would come from the partial destruction of the meson rather than the proton.

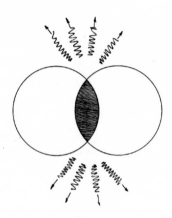

In the enormously high temperatures in the sun protons and neutrons – the nuclei of hydrogen atoms - are travelling at extremely high velocities. Their collisions have sufficient impact to overcome their electric charge repulsion. As they converge captured energy is evicted and they fuse into helium nuclei. This process, called nuclear fusion, also occurs in the hydrogen bomb.

In the hydrogen fusion-bomb the temperatures required to fuse hydrogen nuclei in a couple of litres of heavy hydrogen (hydrogen atoms with one or two neutrons along with the single proton in the nucleus) are generated by the detonation of an atomic fission-bomb.

As the nuclear vortices fuse they lose 0.72% of their mass. The destruction of mass is accompanied by an equivalent release of nuclear energy. Nuclear energy is also released in nuclear fission. In this process a large nucleus, such as that of uranium, splits to form the two smaller nuclei. Nuclear energy is released because protons and neutrons are more tightly bound in the smaller nuclei than in the uranium.

The loss of mass and resultant release of nuclear energy is considerably less in nuclear fission than nuclear fusion. This is because the degree of convergence of nucleons is much less after fission than occurs in fusion; therefore the displacement of nuclear energy is less. Nuclear fusion involves the primary convergence of protons. Nuclear fission involves an increase in degree of convergence of protons and neutrons already bound into atomic nuclei.

The capture of energy, by protons and neutrons, explains the strong nuclear force. When protons and neutrons converge to form an atomic nucleus they become bound together and the more they converge the more tightly bound do these nucleons become. This binding of nucleons — neutrons and protons — in the nucleus of an atom is called the strong nuclear force.

The strong nuclear force can be explained by the vortex. As nucleons converge, the captured energy would begin to swirl between them. The swirling of captured energy, around the centres of the two or more vortices, would act as a force to bind them together.

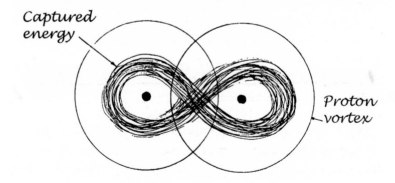

Captured energy

Proton vortex

Because the strong nuclear force originates from captured energy it would operate over a range limited to the diameter of an atomic nucleus. This explains why the strong nuclear force is confined within the atomic nucleus.

When, after nuclear fission, the vortices became more converged, the distance between their centres would decrease. The captured energy would then swirl in a shorter circuit, which would bind them more tightly together. This would be accompanied by a proportionate loss of mass and release of nuclear energy as more captured energy is displaced. Because of this relationship between the loss of mass and increase in binding between nucleons, physicists have a formula which links increased nuclear binding to loss of mass.

In the vortex theory the loss of mass is only indirectly linked to the binding of nuclear particles. The increase in nucleon binding results from the shorter circuit for the captured energy travelling between the two nucleon centres. The associated loss of mass from displacement of captured energy is a separate consequence of the convergence of vortices saturated with captured energy. The link between mass loss and binding appears in nuclear fission and fusion because in these processes only a small percentage of the captured energy is displaced when the nucleons converge. Most of the captured energy remains in the nucleus to bind it together. This more than compensates for the loss.

The vortex account for nuclear energy and the strong nuclear force can be summarised in a hedgehog analogy. The proton vortex could be likened to a hedgehog and captured energy to fleas. Just as every hedgehog carries a resident population of fleas so every proton would carry captured energy. The hedgehog never did anything to acquire its fleas; they just hopped on board. So the proton did nothing to acquire its captured energy; the wave-kinetic energy simply drove into the proton vortex.

If a hedgehog is weighed, the greater part of the measured weight would be that of the hog, but a lesser part would be that of its fleas. In like manner the greater part of the measured mass of a proton would be that of the stable proton vortex, but a lesser part would be that of the unstable swirl of captured energy it contains.

The total flea population would represent the meson particle of captured energy. If the fleas were collected and weighed they would give an indication of the hedgehog's capacity for fleas. In the same way, the mass-energy of a pi–meson would represent the capacity of a proton to contain captured energy. The hedgehog prickles could be taken to represent charge repulsion between protons. Because of their prickles hedgehogs only converge if they are pushed together. So it is that protons will only converge if they collide with considerable force. As the prickles of the two hedgehogs are pushed together there would be less space between them for fleas and so some of the fleas would be evicted. In the same way, as two protons converge there would be less space within them for captured energy and so some would be lost and would radiate away as nuclear energy. With the loss of fleas, the weight of the converged hedgehogs would be less than the sum of their weights before they were brought together. So it is the mass of two converged protons would be less than the sum of their masses before they collided. Most of the fleas would remain on the hedgehogs and not being bothered about which back they bite, would leap from hog to hog. In like manner captured energy, unconcerned, within which vortex it swirls, would circulate between the converged protons.

The formation of neutrons requires an explanation in terms of the vortex model. Neutrons are born in stars as a result of collisions between high-energy electrons and protons. The electron vortex, driven by partially captured energy, could be visualised as smashing into a proton where it becomes enmeshed and held by the force of electric charge acting between them. In the neutron, the proton would immobilise the electron by virtue of its greater static inertia. Meeting the resistance of swirling captured energy within the proton vortex, the electron would come to rest just outside the region occupied by the completely captured energy. It could be pictured as lodging on the surface of the proton represented by the outermost range of the strong nuclear force corresponding to the outermost region of the proton vortex capable of complete capture. The kinetic energy of the electron — its partially captured energy — corresponding to 1.5x its rest mass, would cause it to vibrate. Held to the proton by opposite charge attraction, but driven by its kinetic energy, one could imagine the electron 'bouncing' on the neutron. If one of the bounces were strong enough, the electron would escape from its electric bondage to the proton. The resultant spontaneous explosion of the high-energy electron from the neutron would account for beta-decay, a major aspect of natural radioactivity.

We may understand nuclear energy – the fire of Shiva – but how safe is it for us to deploy it. On November 26, 1977 the early evening television news in South East England was intercepted. Five transmitters were taken over simultaneously and a voice was heard by viewers in London and the South East, from Southampton to Kent. Television engineers were unable to trace the voice, stop it or suppress it and their instruments indicated that normal transmission was in progress. BBC radio host Rex Dutta, published a transcript in his newsletter: Viewpoint Aquarius.

It was 6:12 p.m. Ivor Mills was the newscaster on ITV Southern Television. His face was visible, but his voice was replaced for five and a half minutes with an extraordinary message:

"We speak to you now in peace and wisdom, as we have done to your brothers and sisters all over your planet Earth. We come to warn you of the destiny of your race and your world so that you may communicate to your fellow beings the course you must take to avoid the disasters, which threaten your world and the beings in the worlds around you. This is in order that you may share in the great awakening as the planet passes into the New Age of Aquarius. This new age can be a time of great peace and evolution for your race, but only if your rulers are made aware of the evil forces that can overshadow their judgments.

"Be still now and listen, for your chance may not come again. For many years your scientists, governments and generals have not heeded our warnings. They have continued to experiment with the evil forces of what you call nuclear energy. Atomic bombs can destroy the Earth and the beings of your sis-

ter worlds in a moment. The wastes from atomic power systems will poison your planet for many thousands of years to come. We who have followed the path of evolution for far longer than you have long since realised that atomic energy is always directed against life. It has no peaceful application. Its use and research into its use must be ceased at once or you all risk destruction. All weapons of evil must be removed. The time of conflict is now past. The race, of which you are a part, may proceed to the highest planes of evolution, if you show yourselves worthy to do so.

"You have but a short time to learn to live together in peace and goodwill. Small groups all over the world are learning this and exist to pass on the light of the dawning new age to you all. You are free to accept or reject their teachings, but only those who learn to live in peace will pass to the higher realms of spiritual evolution. Be aware also that there are many false prophets and guides operating in your world. They will suck your energy from you, the energy you call money, putting it to evils ends and will give you worthless dross in return. Your inner divine self will protect you from this. You must learn to be sensitive to the voice within that can tell you what is truth and what is confusion, chaos and untruth.

Learn to listen to the voice of truth which is within you, and you will lead yourselves onto the path of evolution. This is our message to our dear friends. We have watched you growing for many years, as you too have watched our lights in your skies. You know now that we are here, and that there are more beings on and around your Earth than your scientists admit. We are deeply concerned about you and your path towards the light and will do all we can to help you. Have no fear. Seek only to know yourselves and live in harmony with the ways of the Earth. We of the Ashtar Galactic Command thank you for your attention. We are now leaving the planes of your existence. May you be blessed by the supreme love and truth of the Cosmos."

Ivor Mills had a nervous breakdown after the interception of his news broadcast and the Independent Broadcasting Authority dismissed the message as a hoax.

The technology required to overlay a television broadcast in a way that the overlay would not show nor be suppressed by standard broadcasting equipment was not available in the 1970's. It was certainly beyond the capacity of amateur hoaxers. Even if the message were a hoax, that would not detract from the relevance of its content.

Coomaraswamy A. K., *The Dance of Shiva,* Noonday Press, 1969

Dutta Rex, *Flying Saucer Message,* Pelham Books, 1982

Feynman Richard (with Leighton & Sands), *The Feynman Lectures on Physics* Addison Wesley, 1963

Chapter 11

The Unified Field

When the solution is simple, God is answering

Albert Einstein

In the *A Brief History of Time*, Stephen Hawking stipulated: "*A theory is a good theory if it satisfies two requirements: It must accurately describe a large class of observations on the basis of a model that contains only a few arbitrary elements, and it must make definite predictions about the results of future observations.*"

The minimization of assumptions in a theory is called '*Ockham's razor*' named after the 14th century philosopher, William of Ockham who stated: non sunt multiplicanda entia praeter neccessitatem, *i.e.* '*Entities are not to be multiplied beyond necessity*'. Ockham argued that a thesis was more valid for the less arbitrary assumptions it required people to accept. In a contest between theories his standard was simple; accept as true the thesis, which makes least assumptions. He used this law of economy of ideas with such sharpness that it came to be known as Ockham's razor. Galileo invoked the law of economy of ideas, in defending the simplest hypothesis for the heavens. It has been employed by numerous scientists ever since.

A theory is an argument of points derived from a basic set of assumptions. These are called '*axioms*'. It is not necessary to provide experimental evidence in support of the axioms of a hypothesis. The axioms of a theory are not derived from experiment, they are tested by experiment. They often originate from an insight or flash of intuition or they may come from observation of patterns in nature.

The ultimate scientific theory is one that can explain all the particles and forces of nature from a single axiom and account for the entire body of experimental evidence available in science. If it can do so in a way that is understandable to everyone then, according to Stephen Hawking in *A Brief History of Time*, it would be the ultimate triumph of human reason leading to knowledge of the mind of God:

"If we do discover a complete theory, it should in time be understandable in broad principle by everyone, not just a few scientists. Then we should all, philosophers, scientists, and just ordinary people, be able to take part in the discussion of the question of why it is that we and the universe exist. If we find the answer to that, it would be the ultimate triumph of human reason — for then we would know the mind of God."

The Vortex theory is a strong contender. The single idea that subatomic particles are vortices of energy led to an account for:

* **Mass** as the amount of energy spinning in a subatomic vortex.
* **Inertia** as caused by the spin of energy in the spherical vortex.
* **Space** as the infinite extension of the vortex of energy.
* **The curvature of space** is a consequence of space being an extension of mass rather than distorted by mass as speculated by Einstein.
* **3D extension** of space and matter as set up by the 3D form of the spherical vortex of energy.

* **Electric charge** as the interaction of overlapping vortices of energy extending into infinity.
* **Magnetism** as caused by the rotation of vortex particles
* **Time** as one vortex of energy relative to another.
* **Nuclear binding** as the swirl of captured energy between subatomic vortices in the nucleus of an atom.
* **Nuclear** energy as the release of energy captured by subatomic vortices
* **High energy** particles as short lived swirls of energy passing through subatomic vortices
* **Strangeness** as the longevity of a high energy swirl of energy caused by a stable vortex at its centre.
* **Wave-particle duality** originating from couplets of wave trains and vortices of energy
* **Matter and Antimatter** as the symmetry between vortices with equal mass but opposite direction of spin.
* **Gravity** as the interaction between matter and antimatter through the smallest realms of space.
* **Expansion** of the Universe as the interaction between matter and antimatter through the largest realms of space.

Space
Electric charge
Magnetism
Time
Inertia
Mass
Vortex
Gravity
Nuclear binding
Matter & antimatter
Universal expansion

The expansion of the Universe provided a means of testing the vortex theory to satisfy Stephen Hawking's second criteria for a good theory.

Prior to the 1990's it was assumed that the Universe was expanding at a uniform rate. However, in 1990 the physicist Saul Perlmutter of Berkley, California pulled together a team in the USA and UK to look for supernova explosions in distant galaxies. Supported by another team they published results in 1998 from observations of a couple of dozen supernovas that suggested the expansion of the Universe isn't slowing down but is speeding up. Since then more supernova discoveries have supported their conclusions.

The vortex model predicts galaxies are accelerating apart. This accelerating expansion of the Universe is considered to be caused by the electric attraction between matter and antimatter over 'the largest sphere of space'. This was published in 1995 in my book *The New Science of the Spirit* (College of Psychic Studies) three years prior to publication of the discovery of the accelerating expansion of the Universe by the astrophysical teams studying supernova explosions.

The vortex prediction for gravity and universal expansion has been verified by an astronomical observation that was considered in *Science* magazine to be the most important scientific discovery in 1998. As the vortex cosmology is integral to the entire physics, the vortex hypothesis as a whole has been proved by the scientific method and stands as a valid theory in science. This has been achieved by explaining the entire body of physics in terms of a single assumption as well as making a prediction about the accelerating expansion of the Universe which has been proved by subsequent observation.

The vortex explains away all the properties of material

substance. Particles of energy are revealed as forms of pure motion where no substantial thing exists to move. They appear more as thoughts than things. A mind is a body of thought. The Universe, as a body of energetic particles, would therefore, be a mind. Whether you believe the Universe is the mind of God or God is the collective intelligence of the mind we call 'Universe', it is clear that the Vortex theory, which is understandable by everyone, leads us to know the mind of God.

But what of other theories in physics? Where do they fit into the picture?

Ash David: *The New Science of the Spirit,* College of Psychic Studies, 1995

Rees Martin: *Just Six Numbers,* Weidenfield & Nicolson 1999

Hawking Stephen: *A Brief History of Time,* (1988) Bantam Press.

Chapter 12

Uncertainty in Physics

Will some unknown young scientist find a new way of looking at fundamental physics that clarifies the picture and makes today's questions obsolete?

Murray Gell-Mann

It is taught in physics today that the most fundamental particles in matter are quarks. On the 27th January 1977, BBC Television broadcast a report on the new physics in a Programme entitled, *The Key to the Universe*. In the companion book to this Programme, under the same title, Nigel Calder wrote about how the theory for quarks originated:

"*In the early 1930s the contents of the Universe seemed simple. From just three kinds of particles, electrons, protons and neutrons you could make every material object known at the time. Thirty years later human beings were confronted with a bewildering jumble of dozens of heavy, apparently elementary particles, mostly very short lived. They came to light either in the cosmic rays or in experiments with the accelerators. The particles had various mass-energies and differing qualities such as electric charge, spin, lifetimes and so forth. Moreover they were given confusing, mostly Greek names so that one of the most eminent of physicists, Enrico Fermi, was driven to remark before his death in 1954, 'If I could remember the names of all these particles I would have been a botanist'.*

"*The proliferation could be understood, to some extent, in that many of the particles seemed to be energetic relatives of the proton. Because they possessed greater inherent energy their masses were greater. Each was in some sense less tightly bound together than a proton and it could quickly change into a proton with a release of binding energy and an associated loss of mass. But that implied the proton was not a truly basic particle: it was made of something else, which could be bound more or less tightly together.*

"A small group of theorists brought order out of chaos. The principle figure amongst them was Murray Gell-Mann of Caltech (The California Institute of Technology), then in his early thirties. He declared that all the heavy particles of nature were made out of three kinds of quarks. He had the word from a phrase of James Joyce 'Three quarks for Muster Mark.' It was the mocking cry of gulls, which Gell-Mann took as referring to quarts of beer, so he pronounced quark to rhyme with 'stork'. Many other physicists rhymed it with 'Mark'. In German, as skeptics were not slow to notice, 'quark' meant cream cheese or nonsense...."

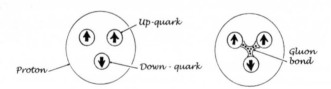

The vortex theory and quark theory are completely incompatible as quark theory is a continuity of the atomic hypothesis of Democritus that there are fundamental, substantial particles in matter with pre-existing qualities.

Quarks exist only in the imagination of theoretical physicists. They do not occur in nature. There are no quarks in the atom because despite decades of research in high energy physics no quark as such has ever been detected. It has been estimated that each of the three quarks in a proton would be five times as massive as the proton itself. Physicists contend that when quarks form a proton they lose fourteen fifteenths of their mass. As the math shows nuclear binding increases with loss of mass scientists contend that quarks would then be so tightly bound they could never be detected in the state. That, they say, is why no one has ever seen a quark!

From the vortex understanding of nuclear energy it is clear that increase in binding with loss of mass cannot occur indefinitely. It only occurs in nuclear fission because a minute amount of mass is converted into energy. The majority of binding mass remains in the protons to bond them together. The decrease in distance between proton centres, that tightens the nuclear bonding, more than compensates for the small loss in captured energy 'glue'. There is no way the bonding would tighten indefinitely with loss of the majority of nuclear binding glue. Richard Feynman warned *"Nuclear energy... we have the formulas for that, but we do not have the fundamental laws."*

Formulae without fundamental understandings lead to fallacy. Richard Feynman was quoted by Nigel Calder as saying; *"The problem of particle masses has been swept in the corner."* He was clearly unhappy with the discrepancy between the mass of the quark and the mass of the proton.

Quark theory incorporates the concept of fractional charge. Up-quarks are supposed to have $\frac{2}{3}$ charge and down-quarks $\frac{1}{3}$ charge. In the proton $\frac{2}{3} + \frac{2}{3} - \frac{1}{3} = 1$ which gave the proton unitary charge. In the neutron $\frac{1}{3} + \frac{1}{3} - \frac{2}{3} = 0$ which gave the neutron zero charge. There is no evidence for fractional charge in nature. All charged particles have unitary charge with is either positive or negative. Charges add up or cancel out by the accumulation of particles with the same or opposite whole unit of charge.

In the vortex hypothesis charge is an effect of the vortex which cannot be fractionalised. Electric charge can only be increased or decreased by adding or subtracting the number of vortices that accumulate at a given point. This is why the charge on a particle is independent of its mass; as observed in experimental physics.

Another fact, which throws the quark theory into question, is the life span of a proton which has been estimated at 10^{33} years, that is a billion, trillion, trillion years whereas one ten billionth of a second is considered a long life span for one of the new particles found in high-energy research. As spontaneous proton decays have never been observed, the proton is infinitely more stable than any of the new, heavy particles that have come to light in the accelerators.

Imagine you walked down a road between two building sites. One offered a trillion year guarantee for their houses whereas the homes on the other side of the street collapsed within a billion of a second of being built. Would you assume the same bricks, cement and construction techniques were employed on both sites? Wouldn't you think there must be something fundamentally different about the buildings on either side of the road to account for the difference in their lifespan?

In the vortex theory the disparity in life span of particles fabricated in the high energy labs and protons is precisely what we would expect from the quantum laws of motion. The short-lived particles have been created out of the massive amounts of energy, forced through stable proton vortices in the high-energy particle accelerators. Energy takes on the vortex form as it is pressed through an atomic nucleus but reverts immediately to the wave form as soon as it emerges on the other side.

The scientists at Fermilab in Illinois, who claimed to have discovered the top quark, didn't actually see a quark. They saw only the tracks of a jet of electrons and muons, which they supposed to be the breakdown of W-particles, which they assumed to be the breakdown products of a top quark. They didn't see a quark because there are no quarks. Quark theory is a fable.

Quark theory has been used to explain the neutron. Scientists say that when a proton interacts with an electron, a force called the weak nuclear force comes into play to transform an up-quark into a down-quark and release an anti-neutrino. In this process the electron and proton have ceased to exist, their place being taken by the neutron. If the process is reversed, then the proton and electron come back into existence and a neutrino is released. If there are no quarks this account for the neutron falls over and along with it, the concept of a weak nuclear force.

Neutrons can be formed out of an electron and proton. The neutron has the sum mass of an electron and proton. Outside of an atomic nucleus, after a few minutes, a neutron will fall apart into an electron and a proton. This suggests the neutron is a bound state of electron and proton. The presence of charged particles in the neutron is supported by its 'magnetic moment'. The neutron, if it were a truly neutral particle, would have no magnetic moment because the magnetism of a particle is created by the spin of its charge. The magnetism of a neutron adds support to the view that it is a bound state of two opposite charges, which mostly cancel each other out rather than a single particle with no charge at all. So why don't physicists accept the neutron as an electron bound to a proton as all the evidence suggests?

The reason why scientists are adamant that a neutron is not an electron bound to a proton is because accepting that could signal the end of quantum mechanics!

A major pillar of quantum theory is the uncertainty principle proposed, in 1927, by the German physicist Werner Heisenberg. He suggested that attempting to observe things in the sub-atomic world increases the uncertainty about what is going on there. If you want to look at a particle in order to ascertain its position, you would have to reflect light off it. But, the act of bouncing light off the particle would give it a kick that would increase its momentum and so make its position more uncertain.

Looking at small objects requires more energy than is required for looking at large ones. This is evident in the electron microscope, which employs higher frequency radiation than a light microscope. Because sub-atomic particles are the smallest things in nature, the action of ascertaining their position with any degree of certainty would require a very great energy, which would give them an enormous kick. Heisenberg argued against being able to determine, with any certainty, both the position and the momentum of a particle. This came to be known as the 'Heisenberg uncertainty principle'.

If a neutron is an electron bound to a proton, the position of the electron would be known for certain; somewhere within the circumference of a proton. If Heisenberg's principle is applied, the high degree of certainty in position demands an enormous uncertainty in momentum. Electrons in the neutron would have velocities up to 99.97% of the velocity of light. How could electrons be locked in neutron, as the evidence suggests, if they are racing round at the velocity of light? If they were they would have a mass up to forty times the mass of an electron at rest. This would show up in the measured mass of neutrons and it doesn't.

Albert Einstein despised Heisenberg's principle. He described it as, *"...a real witches calculus...most ingenious, and adequately protected by its great complexity against being proved wrong."*

The uncertainty principle is a block to unifying relativity and quantum theory. As Stephen Hawking commented, *"The main difficulty in finding a theory that unifies gravity with the other forces is that general relativity is a classical theory in that it does not incorporate the uncertainty principle of quantum mechanics."*

All the evidence suggests a neutron is an electron bound to a proton. The fact that Heisenberg's formula doesn't fit implies the problem lies with Heisenberg's principle not the neutron. As Richard Feynman said, *"If your theories and mathematics do not match up to the experiments then they are wrong."*

Scientists may protest that Heisenberg's principle must be right because of its incredible success in physics. However, the successful application of a principle does not prove it is valid. A car may run well but if it fails its roadworthy test it has to be scrapped. The neutron, discovered five years after the uncertainty principle was proposed, was the experimental test for the principle. It failed the test and should have been scrapped long ago as Einstein suggested.

Instead Einstein was scrapped! Unwilling to accept improbability and uncertainty he bowed out of mainstream physics. The fellows he left behind went on to build quantum mechanics on the back of the uncertainty principle. Their theories for force came to depend upon it. If the principle is disproved, the most successful theories in physics collapse, taking the scientific method with them. That is why scientists vehemently deny that the neutron is an electron-bound to a proton.

Quantum mechanics came to depend upon the uncertainty principle after physicists suggested particles could borrow energy from the Universe to bring about the creation of short-lived force-carrying particles. So long as these 'virtual' force-carrying particles were sufficiently unstable to decay and repay the energy debt within the time allotted by Heisenberg's formula then no conservation law would have been broken in the overall process of their formation and decay. If the time span of their existence was short enough, the uncertainty principle allowed for very large amounts of energy to be involved in the formation of the particles, which would, in turn, enable them to carry very powerful forces.

The underlying premise was that 'anything is possible

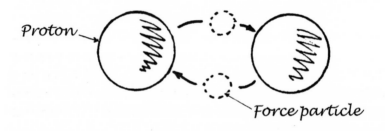

behind the screen of uncertainty'. The idea, that sub-atomic particles run credit with the Universe, as though it were a bank, is speculative but because it is impossible to be certain that it doesn't happen, no one could argue against it.

The Talmud states: *'We see the world, not the way it is, but the way we are'.* This law of projection applies in the way that the scientist perceives the 'quantum world'. With his mortgage, bank loan, and innumerable credit cards, the professional physicist — like most other people in the modern world — runs his affairs on credit. It is this image that has been projected onto the sub-atomic world. The proton is imagined to borrow energy in order to create gluons for the binding of quarks, W and Z particles for the weak nuclear force, mesons for the strong nuclear force, virtual-photons for the forces of electric charge and magnetism and gravitons for the force of gravity.

The vortex account for forces does not require the as-

Mr Proton

sumption of anthropic processes going on inside protons, hidden behind a shroud of uncertainty. Neither does it require that protons take any action, such as borrowing energy, in the formation of force carrying particles. Forces are a natural result of the dynamic nature and structure of the vortex.

Physicists have accepted the energy credit system of quantum mechanics because the theory has been spectacularly successful in predicting the outcome of experiments. Whilst it cannot be denied that many theoretical predictions have led to major discoveries in physics there is great danger in putting too much store by predictions. The spectacular vindication of a prediction can cause a sensation that is liable to give more credence to a theory than it is worth. Why should a theory be treated as true if it makes a successful prediction? There can be a number of different explanations for a scientific experiment or observation. It might work as well and not require belief in speculative absurdities such as primary particles running lines of credit with the Universe.

However, appearing as an explanation rather than a prediction, alternative accounts may pass unnoticed, because predictions drive science. Predictions suggest directions for research and designs for experiments.

Some predictions, incorporating the uncertainty principle, have been gloriously vindicated by experiment. Richard Feynman's theory for electro-magnetism is the most successful theory in the history of science. If the neutron disproves the uncertainty principle it would collapse his 'quantum electro-dynamics' and take the scientific method with it which would undermine the very foundations of science. Then it would be clear that scientists have proved what they believe to be true rather than what is true. That would confirm that we create our own reality.

Calder Nigel, *Key to the Universe: A Report on the New Physics* (1977), BBC Publications.

Hawking Stephen, *A Brief History of Time* (1988) Bantam Press.

Hawking Stephen, Black Holes and Baby Universes, Bantam 1993

Feynman Richard with Leighton & Sands, The Feynman

Lectures on Physics, Addison Wesley 1963

Matthews R, Unravelling the Mind of God, Virgin Books 1992

Chapter 13

The Power of Uncertainty

Even for the physicist the description in plain language will be a criterion of the degree of understanding that has been reached.

Werner Heisenberg,

In his inaugural lecture, Stephen Hawking said that because of the Heisenberg uncertainty principle, electrons could not be at rest in the nucleus of an atom. All the evidence suggests that they are. If the uncertainty formula doesn't match what we know about the neutron from experimental physics then Heisenberg's theory is wrong. The uncertainty principle is a major pillar of quantum theory and the foundation of quantum mechanics. If it is wrong, where does that leave quantum theory?

In his book 'The Problems of Physics', Prof A.J. Leggett wrote: *"Quantum mechanics...has had a success which is almost impossible to exaggerate. It is the basis of just about everything we claim to understand in atomic and sub-atomic physics, most things in condensed-matter physics, and to an increasing extent much of cosmology. For the majority of practicing physicists today it is the correct description of nature, and they find it difficult to conceive that any current or future problem of physics will be solved in other than quantum mechanical terms. Yet despite all the successes, there is a persistent and, to their colleagues, sometimes irritating minority who feel that as a complete theory of the Universe, quantum mechanics has feet of clay, indeed 'carries within it the seeds of its own destruction'."*

In his book 'Unravelling the Mind of God', Robert Matthews said: *"The Copenhagen interpretation of quantum theory remains the most widely accepted way of looking at quantum theory amongst scientists today, but it should be stressed that, despite what some might claim, it remains just that: an interpretation. There is powerful experimental evidence that it is an acceptable way of looking at the world, but it is most definitely not the only possible interpretation."*

Heinz Pagels wrote, *"Something inside us doesn't want to understand quantum reality. Intellectually we accept it because it is mathematically consistent and agrees brilliantly with experiment. And yet the mind is not able to rest."*

Quantum theory is virtually impenetrable even to the greatest minds. As Richard Feynman once remarked, *"Quantum theory...you never understand it, you just get used to it."*

Lord Rutherford, the father of nuclear physics said, *"These fundamental things have got to be simple."*

There is something fundamentally wrong with quantum physics that it is so complex and difficult to understand. The quantum theory for forces has been immensely successful. Though it is logically absurd that particles run lines of credit with the Universe to operate forces, mathematically the theory is a masterpiece. Physicists love it. But it does depend upon the uncertainty principle and that is its weakness.

The problem for physics today is that the theories for forces have been vindicated by the scientific method. If the quantum theory for forces goes it takes the scientific method with it. With failure of the uncertainty principle we are confronted with an end to science as we know it.

One way to save the uncertainty principle is to say it

does not apply to neutrons. Electron — proton couples, in which electrons orbit protons, are already excluded from the principle. Particles moving in orbits, under the inverse square law of attraction, are excluded from the uncertainty principle because the line spectra of atoms, produced by changes of electron orbits are exact. Some physicists may be tempted to treat the neutron as a proton with an electron in a very tight high-energy orbit so that it can also be excluded from Heisenberg's principle.

However, in 1956 the American physicist Chien Shiung Wu lined up the nuclei of radioactive atoms in a magnetic field so they were all spinning in the same direction and observed that more electrons were emitted in one direction than in another. This experiment showed that electrons appear to emerge from a specific site on a neutron rather than being randomly distributed in a high speed orbit.

It would be more honest for scientists to admit that the line spectra of atoms back up the evidence of the neutron to prove the uncertainty principle has failed. It is time for science to face up to the truth. The uncertainty principle has to be scrapped and along with it the current theories for forces. With the advent of a new millennium it is timely to make a fresh start in physics.

On April 29, 1980, Stephen Hawking was inaugurated as Lucasian Professor of Mathematics at Cambridge. In his Inaugural Lecture entitled, 'Is the end in sight for theoretical physics?' he discussed the possibility that the goal of theoretical physics might be achieved in the not too-distant future: say by the end of the 20th century. Many scientists believe we are in sight of a complete understanding of the Universe and no new major discoveries should be expected in future physics. The same view was expressed at the end of the 19th Century. Except for two small clouds on the horizon, scientists thought they had solved all the problems. No new basic discoveries were expected and the future work in physics was believed to lie in improved methods of measurement to the sixth decimal place. The two small clouds grew into hurricanes, which swept away the theories of 19th Century physics and led to quantum theory and the theories of relativity. History repeats itself. Once again clouds have appeared on the scientific horizon, which threaten a storm that could blow away the major theories of 20th century physics.

However, there is no future for science if experimental observations are dismissed in order to save the theories. In the words of George Gamow, *"Staggering contradictions of this kind, between theoretical expectations on the one side and observational facts or even common sense on the other are the main factors in the development of science."* Abandoning the old theories of the 20th century may enable us to adopt a simple, understandable and unified approach to physics in the 21st century which will lead and a completely new understanding of the Universe.

Failure of Heisenberg's principle would invalidate the quantum theories for force. This in turn could reveal the shortcomings of the scientific method but paradoxically this could turn into a real triumph as it would maintain uncertainty about theories and keep the door open for future generations to continually reinterpret science and uphold the premise of Popper.

The Austrian philosopher of science, Karl Popper warned that the scientific method should not be used to prove theories. He used an analogy of black and white swans to stress his point that science is not in the business of proving theories but rather of disproving them.

Someone could have a theory that all swans are white

but even if a thousand white swans were counted the belief would not be proved and the addition of more white swans wouldn't make it any truer. However, the appearance of a single black swan would disprove the theory altogether. Scientific theories are white swan belief systems, which survive until they are destroyed by the appearance of a black swan, that is, the arrival of even a single fact, which makes it clear that they cannot be true. However, it would be naive to imagine that theorists would make predictions in order that their theories can be disproved and themselves discredited. It is human nature to aim for success, not failure. In the real world, theorists live in hope that experiments will uphold their predictions and pave the way for a Nobel Prize

By revealing the shortcoming of the scientific method Heisenberg's principle of uncertainty has released humanity from the burden of truth imposed by science. In the past religions dictated truth, telling people, on pain of death, what they were to believe. Today there is a tendency for scientists to assume they have the true understanding of the Universe and heap scorn on alternative worldviews.

Thanks to Werner Heisenberg this is quelled and uncertainty remains sacrosanct in science. We can put forward theories but we can never be certain that they are true. As waves breaking endlessly on the shore so generations can throw their interpretations of truth on the beach of uncertainty and no one can say for certain, '*this is the way it is, we are right and you are wrong*'.

In quantum theory it is believed that nothing exists until it is observed. The idea that reality depends on observation is profound. Consider yourself. Who are you? Are you your body or your mind and emotions or are you the conscious awareness that observes your body, mind and emotions? If quantum theory is correct your body, mind and emotions depend on conscious awareness for their existence and if there is no separation between human consciousness and cosmic consciousness then collectively we create our own reality.

The scientific method is subject to this law. Observations are liable to conform to collective human belief. Uncertainty is a humble acknowledgement that we cannot be certain of anything because quantum reality can mould around our certainty.

As the conscious observer, beyond the limitations of physical space and time we would have infinite knowledge and unlimited potential so the whole value of our human life would lie in the uncertainty of the human situation. Accept that and everything that happens to us would be seen as perfect for our growth and development. We could then move beyond judgement into love. Uncertainty may be a challenge but it can also be a gift.

In science uncertainty has levelled the playing field. We cannot be absolutely certain quarks and force-carrying particles do or do not exist and the same applies to angels. Throughout history people have believed in angels and today they are as popular as ever. Surprisingly there is a precedent for the angels in Einstein's theory of relativity.

Feynman Richard with Leighton & Sands, The Feynman
Lectures on Physics, Addison Wesley 1963

Gamow George, *Thirty Years that Shook Physics*, Heinemann.

Hawking Stephen, Black Holes and Baby Universes, Bantam 1993

Leggett A.J. *The Problems of Physics*, Oxford University Press, 1987

Matthews R, Unravelling the Mind of God, Virgin Books 1992

Pagels Heinz, The Quantum Code, Michael Joseph 1982

Popper, Karl, The Logic of Scientific Discovery, Hutchinson 1968

Chapter 14

Beyond the Speed of Light

The mouse felt that beyond the world of mice another higher deeper inner reality transcended all gnawing squeaking nibbling ways of the world like one colossal cheese awesome yellowish with lovely holes irrefutably sublime

Alfred Brendel

Before the turn of the 20th century, mass, space, time and the laws of physics were thought to be absolute and invariable. However, in the course of the 20th century, it became clear that these classical assumptions were invalid. Einstein showed that space and time, mass and the laws of physics were not absolute and invariable but were relative to the speed of light. Since then, most scientists have proclaimed that there is nothing absolute in the world of physics, but still, throughout the 20th Century, the speed of light was generally accepted as absolute and invariable — the ultimate frontier of the universe of energy. Now at the turn of the 21st century the time has come to challenge this final bastion of absolutism.

Sub-atomic particles cannot be accelerated beyond the speed of light. This is because they are accelerated by wave-particles of energy moving at the speed of light. If a moving particle is a vortex propelled by a wave train of energy, it accelerates when it receives an input of additional quantum units of energy. If the propelling packets

138

of energy are particles of movement at the speed of light then obviously this would be the maximum speed any particles propelled by them could ever achieve. However, this does not mean that the speed of light is absolute. Nor does it prove the speed of light to be the limit of velocity in the Universe. Neither does this mean that all energy is constrained by the speed of light.

Most people in medieval Europe believed in the 'absolute truths' of dogmatic religion. They also believed that the Earth was flat. They were constrained to Europe by their limiting beliefs. Before Christopher Columbus could set out to discover America, he had to break free of the flat earth mentality and see the 'end of the world' as a horizon beckoning him toward the discovery of new worlds.

If we treat the speed of light as the absolute limit to energy, then, like the medieval masses of unimaginative flat-earth thinkers, we will be stranded on the shores of our perceived reality. The process of discovery begins with the entertainment of possibility. Columbus discovered America only because he entertained the possibility that the Earth was not flat, but was a globe on which the east could be reached by sailing west.

It is clear to most physicists today that the speed of light can be exceeded which suggests there is energy beyond the speed of light. Though it may be beyond our immediate perception it need not remain forever beyond the bounds of knowledge because new discoveries are being made every day and new discoveries in physics usually follow theoretical predictions.

To make progress in science it is necessary to make imaginative postulates and then test them against a body of collected information. As Karl Popper said, *"If you want to promote the growth of scientific knowledge, you should adopt the method of extravagant guess followed by unrestrained criticism in which the guesses found to be false are cast from the body of science."*

The guess that energy exists beyond the speed of light is extravagant but there are many things yet unexplained that could be explained if we allow for this possibility. Energy beyond light could be called super-energy. If super-energy does exist we need to consider the way it relates to us and how it affects our view of the Universe.

There is tradition in science of basing predictions as closely as possible on what is already observed. The most successful predictions in the history of science are the ones that have followed this principle of symmetry – a tradition originating in the ancient Hermetic principle: *As above so below, as below so above.*

When making predictions in science it is important to restrict variables and reduce speculation to a minimum so economy of ideas as well as symmetry is important. Prediction is, by its very nature, speculative so economy in the elements of the prediction helps to reduce the degree of speculation. Speculation can also be restrained by avoiding wide deviations from what is already observed. The most successful predictions in science have been based on observations in nature.

From the *As above...* symmetry of Hermes it can be predicted that **the universe is like the atom**. This symmetry that the whole is represented in the parts is the fractal or hologram principle. The fractal is fundamental to the structure of the Universe. It is central to Chaos theory and is illustrated by the Mandelbrot set.

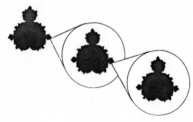

In 1979, on a computer, Benoit Mandelbrot discovered a method of feeding the answer to an equation back into the equation to produce an endlessly repeating pattern. Appearing as an intricate order of discs surrounded by 'galactic' swirls and spirals this, now famous pattern, was named after him. The central discs of the Mandelbrot set are edged with minute projections. When these are expanded they are found to contain repeats of the original pattern. This fractal has the potential to repeat itself endlessly.

Millennia before Mandelbrot, Hermes recognised that we live in a fractal Universe where symmetry applies at every level, to the human, the atom and the Universe. Understanding the atom should lead us to an understanding of the Universe. By understanding the Universe and the atom, we should better understand ourselves.

The atom is organised in discrete quantum levels. If the Universe is organised in defined levels of energy we would be able to apply another of the ancient Hermetic teachings e.g. the Universe is organised in discrete planes — the word 'planet' comes from this idea. Pythagoras developed this idea into the 'Harmony of the Spheres'.

The Hermetic idea of the planes was used by Aristotle in the fourth century B.C.E. and developed by Ptolemy in the second century C.E. into a cosmology that was taught as doctrine by the Catholic Church until displaced by Galileo in 1609.

Aristotle misinterpreted the Hermetic idea by suggesting the Earth is central to the Universe. Though he said it was a globe he taught that the sun, moon and planets orbited the Earth in perfect spheres, and that a starry sphere enclosed the entire system.

Galileo contradicted Aristotle. He embraced the idea proposed by Nicholas Copernicus in 1514, that the sun was central and orbited by the Earth and planets. Johannes Kepler hypothesised that the orbits were elliptical rather than circular. Isaac Newton confirmed his idea in 1687.

Because of Aristotle's erroneous speculation the Hermetic planes have been discredited. However, this is an invaluable model when applied to the idea of different levels of quantum reality in the Universe rather than to planetary orbits. The concept of 'levels of quantum reality', suggest the planes of reality in the Universe are based on different states of energy, much like the quantum states in the atom.

The innermost electron orbits in the atom take the form of concentric spheres. This is same model represents the subatomic particle as concentric spheres of energy.

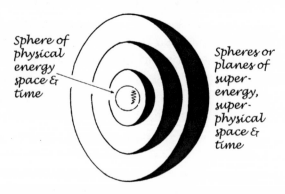

Sphere of physical energy space & time

Spheres or planes of super-energy, super-physical space & time

To minimize speculations, it can be suggested that particles of super-energy occur in the forms of wave and vortex. These could set up atoms and molecules, life forms, planets and stars no different to the way that physical-energy creates the atoms, molecules and the bodies of matter that make up our world. The only difference between them would be the intrinsic speed of their energy.

In our world the intrinsic speed of energy is the speed

of light. This is the '*Einsteinian constant*'. It is this constant that would be different on each plane or level of quantum reality.

We know from the work of Albert Einstein that mass, space and time in our world of physical energy are relative to the speed of light. In the worlds of super-energy, mass, space and time would be relative to higher speeds.

Physical-energy in the vortex sets up the space and time in our reality. In the same way vortices of super-energy would set up space and time in the realities predicted beyond the speed of light. Super-energy could establish other worlds, other domains of space and time quite distinct from our own. The physical world may be but one level of energy in an 'ascending series of quantum realities'.

The prediction of an ascending series of quantum realities in the realms of super-energy accords with quantum thinking. Whereas in classical thinking, it was believed that energy existed in a continuous band, which could be likened to a ramp, quantum theory established the distribution of energy more as a flight of steps than a ramp. The great achievement of quantum theory was to show that energy is discontinuous.

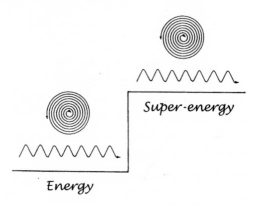

Super-energy

Energy

Electrons in the atom can only exist within certain energy levels — none occur in between. Particles of matter occur with specific mass-energies and none exist between these quantum steps. Just as no one knows why a proton happens to have a mass-energy 1836 times that of an electron, there is no immediate explanation for why the speed of light is critical to physical-energy.

Everything in our world is relative to the speed of light. Other worlds would exist on planes relative to different critical speeds of energy. Our relationship to them would be determined by two simple laws:

1 **The Law of Simultaneous Existence.**

If worlds of super-energy have their own space-time set up by vortices then the dimensions of space and time would not separate the worlds of energy and super-energy. Because vortex motion does not exist between the planes, space and time would exist *in* the worlds, not *between* the worlds. As universal realities are not separated by space and time it follows that:

Universal quantum realities are in simultaneous existence.

2 **The Law of Subsets.**

The dimension that separates the worlds of physical-energy and super-energy is the intrinsic speed of the energy. Lesser speeds of motion are a part of greater speeds of motion. A jet can move at running speed then the speed of a bicycle as it taxies from the terminal. However fast you run or even on your bike there is no way you could match the velocity of the jet as it takes off. Running and bicycle speeds are sub-sets of jet speeds but not the other way round

Worlds based on lesser speeds of energy would be a part of worlds based on greater speeds.

The world of physical–energy is a subset of the world of super-energy.

144

The simultaneous existence of the worlds leads to fascinating possibilities. For example, we may look through our telescopes at a planet that appears to be dead but it may only be devoid of life at the level on which we live. On that planet, at another level of quantum reality there could be a world of super-energy teeming with life. There could be other worlds full of life on and in our own planet. We would not be aware of them because their speed level of energy would be different from our own.

If there were beings in our own world living in their own space and time as we live in ours we would not see them. This is because physical light would not interact with their super-physical bodies and we would not feel them or bump into them as the forces associated with physical matter would not interact with the force fields of their super-physical matter. They would not be apparent to us because they occur outside of our physical space and time. However, they could see us even though we cannot see them if our world of physical energy were a sub-set of their world of super-energy.

In the dimension of speed, the speed of light represents the demarcation between physical-energy and super-energy, between physical reality and the super-physical realities. As such I describe the speed of light as the 'light-barrier'. The light barrier separates the world of physical-energy from the world of super-energy, the physical from the super-physical but not the other way round. Being a sub-set of super-energy, the world of physical-energy would not be apparent as a separate reality from beyond the light barrier.

All speeds are measured from zero. This could be described as the common ground point of all movements. If the Universe consists of a series of quantum realities based upon different critical speeds of movement, then because they all share the same ground point, they would be concentric about this point. The structure of the Universe, as an ascending series of quantum realities would be identical to the structure of the sub-atomic vortex, as a system of concentric spheres of energy. This is an example of the *as above...* symmetry.

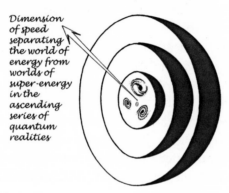

Dimension of speed separating the world of energy from worlds of super-energy in the ascending series of quantum realities

Any thing that occurs in the lesser occurs in the greater reality because the lesser is part of the greater. This is depicted in the concentric sphere model of the Universe. Our physical world is a sub-set of the super-physical. We are intelligent life forms in the physical realm of the Universe so if there is a super-physical realm there would be intelligent life in it. If worlds of super-energy exist, it is possible they would be populated by intelligent beings a lot smarter than us.

As intelligent beings, we would be a part of the world of super-energy. In this context it would be illogical to assume intelligence is confined to a single creature in a lesser part of the whole; that human intelligence is the only manifestation of intelligence in the Universe. We may be the highest level of intelligence in our world but big fish in a little pond are not especially the biggest fish in existence.

The laws of simultaneous existence and subsets suggest that intelligent life forms in the realms of super-energy could see us though we couldn't see them. This is because physical space and time would be a part of super-physical space and time.

Physical-energy would not interact with super-physical bodies because super-physical space and time are beyond physical space and time. This is clear from the *'Mincowski diagram'*.

Mincowski was the mathematician who taught Einstein his math. Working with Einstein on the theory of relativity, Mincowski constructed a diagram showing that speeds beyond the speed of light could not be contained or perceived in the space-time relative to the speed of light.

Existing outside physical space and time, super-physical reality would not be a part of our world, nor subject to its physical laws. However, super-energy could interact with physical bodies because physical bodies occur within the overall domain of super-physical space and time. So whereas super-physical light could reflect off physical objects, physical light would pass right through super-physical objects. It follows that we would be visible in the super-physical world, but the super-physical world would be invisible to us.

Since time immemorial people have believed in the existence of spirits such as angels. In religious tradition angels are believed to be all around us and aware of us. However, in the folk lore we are not normally aware of them because ...*they are moving too fast for us to perceive.*

Was it not in the folk lore: *You do not see the fairies because you are too slow and clumsy!*

Fascinating that the fairy tales recognised that the separation between mankind and elementals was a factor of speed!

The mythology of angels, fairies and other 'spirits',

that their speed bars them from common perception in physical space and time, found credence in the work of Einstein and Mincowski as they developed the theory of relativity. The Mincowski diagram makes it clear that speeds beyond the speed of light cannot be contained in a space time continuum relative to the speed of light. Any being formed of energy moving faster than the speed of light could not be contained in physical space and time. Common light would pass right through them so they would be invisible. Forces would not interact with them so they would not be bound by gravity nor would they be blocked by matter. They could float, walk through walls and pass right through us as we and the walls are mainly space. The only thing that bars us from passing through walls is the force of repulsion between electrons in the atoms in our bodies and atoms in the wall.

One thing is clear from the Mincowski diagram; there is more reason to believe in angels and fairies than in quarks and the virtual particles of quantum mechanics!

The Mincowski diagram also alludes to the principle of simultaneous existence. To quote Gary Zukav from the 'Dancing Wu Li Masters': "*Minkowski's mathematical explorations of space and time were both revolutionary and fascinating. Out of them came a simple diagram of space-time showing the mathematical relationship of the past, present, and the future. Of the wealth of information contained in this diagram, the most striking is that all of the past and all of the future, for each individual, meet and forever meet, at one single point, now. Furthermore, the now of each individual is specifically located and will never be found in any other place, than here.*"

The principle of simultaneous existence suggests that because the forms of energy on each plane set up space and time, all beings existing on all planes of quantum reality in the Universe must exist simultaneously in the same eternal point of 'here and now'. External from each plane, in the eternity of consciousness which brings them into existence by the power of imagination, there is no space and there is no time. There is only the unity point of the immediate existence of everything.

Brendel Alfred, One Finger Too Many, *Faber & Faber*, 1998

Popper, Karl, The Logic of Scientific Discovery, *Hutchinson* (1968)

Zukav Gary, The Dancing Wu Li Masters, Rider 1979

Chapter 15

Evidence in the Corn

When you have eliminated the impossible, whatever re-
mains, however improbable, must be the truth

Sherlock Holmes

You may ask, 'If our world is part of a world of super-energy occupied by intelligent beings that are aware of us, why don't they make themselves known to us?' The answer is they do only we dismiss the evidence they leave for us as hoax. This attitude can be depicted by a 'matchbox analogy'. Imagine a matchbox containing matches that were aware of their surrounds but the limits of their perception are the walls of the matchbox. The match people are unaware of the room beyond the matchbox or the people in the room whereas the people in the room are aware of the match people in their match-box because the matchbox is a subset of the room.

Constrained by the limitations of their perceived real-ity some of the little matchbox men are sceptics. Unaware of intelligent beings in a greater reality they deny their existence. Some of the match people sense there is a greater reality occupied by other people beyond the matchbox but believe they fly with wings like the swan depicted on the matchbox. This causes the sceptics amongst them to scoff.

The people in the room project patterns on the walls of the matchbox to provide evidence of their existence. The response of the sceptics and the majority with match head mentality is to cry *'Hoax'*.

For decades thousands of strange circles have been seen in crops, mainly in Wiltshire in England but increasingly elsewhere in the UK and abroad. Since 1990 the circles changed into pictograms of enormous size and exquisite artistry. Appearing by the score each season, the breathtaking beauty and unbelievable complexity of some of the formations stagger belief. Nonetheless all crop formations have been discredited in the public mind by the claim of hoax.

There is solid scientific evidence to refute the hoax hypothesis for crop formations. It only takes one crop formation to defy the hoax hypothesis and raise the question of the origin of these remarkable pictograms that appear in crops, world wide, year after year. One such appeared in Cherhill, Wiltshire, England in August 1993. The *Journal of Scientific Exploration* published a paper by Prof. Levengood in 1995 which reported:

'...the unusual discovery of a natural iron "glaze" composed of fused particles of meteoritic origin, concentrated entirely within a crop formation in England, appearing shortly after the intense Perseid meteor shower in August 1993...Abnormalities in seedling growth was also consistent with the unusual responses of seeds taken from numerous crop formations...Presence of meteoric material adhering to both soil and plant tissues, casts considerable doubt on this being an artificially prepared or "hoaxed" formation.'

In 1994, in a journal of plant physiology, Dr. Levengood reported anatomical anomalies found only in plants taken from crop formations. This is consistent with the findings of other researchers. The formation of plant tissue when examined under the microscope is very different in plants taken from inexplicable formations against samples taken from hoaxed or artificial formations. In hoaxed formations plant stems are buckled or snapped as the crop is trampled. In the unexplained formations the stalks are 'moulded' into the bent position with no sign of trampling or trespass on the crop. In the unexplained formation the plants are undamaged and continue to grow.

As a scientist, Levengood has scrutinised crop formations worldwide for years and has documented fascinating anomalies in unexplained patterns, not only in the anatomical and physiological changes in the plants but in the soil surrounding them. He noticed microscopic fused particles of iron unique to crop formations; much like grape shot. From the changes in the plant stems − the 'nodes' where bending occurs − and these 'melted' particles in the soil, he concluded that very large amounts of energy have been involved in the formations; energy somewhat akin to microwave radiation.

What Levengood discovered was that the changes in plants - notably node lengths - and concentration of fused particles of iron, diminished in proportion to distance from the centre of the pattern. This suggested the super-imposition of an intense energy field over the crop with intensity falling off with distance from the formation.

Levengood's observations are incongruent with the claims of hoax; especially considering the discovery of node changes in the standing plants as well as the bent plants in the unexplained formations. The science points to a complex, chaotic, thermodynamic energy field of enormous intensity, with components acting unpredictably and independently. It would require advanced electromagnetic technology to bring about the changes in crops and underlying soil. This absolutely defies the simplistic view of hoaxers with ropes and boards.

Crop formations cannot be explained away by science or easily dismissed so they have been discredited by the production of a few fraudulent circles. The media have supported the misconception that all crop formations are a hoax.

Considering the thousands of pictograms that have appeared over the years, the picture of 'Doug and Dave' stumbling out of the pub to create enormous crop formations, throughout the growing season, night after night, often in wild, wet, windy weather is beyond the bounds of credibility. The idea that an army of hoaxers suddenly took it on to 'party in the corn-fields' is ridiculous. People are not that consistent or devoted to maintain the level of hoaxing required accounting for all the crop formations that have been reported and photographed. Also parties of hoaxers could not work the fields, night after night, creating vast, complex, perfect geometric formation in pitch dark without ever being detected and prosecuted for trespass and deliberate damage to valuable crops. People park vehicles and make noises, they use torches. They could never go out thousands of times, with their boards and ropes in the English countryside, and never be caught red-handed. And only the military have access to the sort of technology required to produce massive electro-magnetic fields required to account for the anomalies observed by researchers. Military helicopters have never been noticed hovering over corn crops prior to the appearance of formations. Nor is it likely the MOD would sanction the use of military aircraft to mould patterns in fields of barley, rape and wheat.

In recent years despite the increase in size, complexity and dispersal of crop formations in England and abroad no effort has been made by the mainstream scientific community apart from a few intrepid researchers, like Levengood, to distinguish true from false anomalies and efforts by these individual scientists to show that most crop formations could not be produced by hoaxers, have been ignored.

It is fascinating that crop formations transformed from simple circles that could be duplicated by hoaxers into stunning pictograms that could not be hoaxed in the 1990's just when the 'hoaxing' became generally accepted as the solution to the crop circle enigma.

To quote John Mitchell from the introduction to Nick Kollerstrom's book on crop circles:

After some twenty years of crop circle research no one yet has any idea of what is going on. Every season new and better designs appear in the cornfields. They are amazingly subtle and beautiful. Nothing in the world of art today has anything like their quality…In the early days it seemed plausible that the circles were caused by freak whirlwinds or some other weather effect. That idea became impossible after 1990 when the first elaborately-designed 'pictograms' appeared. These, obviously, were products of intelligent minds. So the theorists were divided. UFO enthusiasts believed the intelligent source to be extra-terrestrial, while most other people took the down-to-earth view that it was all a hoax.

The 'hoax' theory implies that unknown teams of skilled and dedicated artists are secretly at work during the summer nights, stamping or raking out large scale patterns and leaving no evidence behind. That seems the only rational solution. Yet there are so many difficulties in it that experienced researchers are sceptical. No one ever detects these supposed circle-makers, or their cars or equipment. Certain fields are watched, yet circles suddenly appear in them, and nothing has been seen or heard.

Then there is the problem of how these large, complicated patterns could possibly be completed in the few hours of summer darkness, never left unfinished and never with awkward errors. Copyists have been commissioned to make their own circles, legally, in daylight and with no time limit. But none of these has anything like the quality of the great, unclaimed masterpieces that appear spontaneously.

Nick Kollerstrom concluded his book:

Professor Gerald Hawkins spoke to the 1998 meeting of the American Astronomical Society at Washington D.C. on the subject 'From Euclid to Ptolemy in English Crop Circles'. He received a standing ovation. Hawkins had been seeking what he called the 'intellectual profile' of the unknown artists: '...the mechanics of how the crop patterns are formed is a mystery,' he concluded, 'but the intellectual profile behind it all has turned out to be an even greater mystery.'

Lucy Pringle concluded her book on Crop Circles:
Crop circles have wit, beauty, elegance, rhythm and form. Perhaps they are a part of some great imaginative project. The sheer beauty and complexity of crop formations leaves little doubt that an intelligence is at work...

Crop formations validate the law of simultaneous existence and the law of subsets. They provide circumstantial evidence of the existence of beings with intelligence and a highly advanced technology in a world that permeates our own. We are not aware of the intelligence because we see only the pictograms, not the source. It is obvious that the intelligence is aware of our world as it uses fields of crops as canvasses for executing exquisite art.

The obvious non-human intelligence and advanced unseen technology behind crop formations suggests that mind exists in a greater reality than we perceive. Massive pictograms of incredible complexity, impeccable accuracy, exquisite artistry appearing in crops within minutes in dead of night and terrible weather, not to mention uneven ground and no sign of trespass or hoaxer activity suggest the existence of sophisticated, non-physical beings with a love of art and geometry.

Perhaps patterns appearing in the corn fields have something in common with patterns of thought appearing in our brains. Maybe both are the same phenomenon manifesting at a different level; intelligence operating through super energy to bring about changes in the subset world of energy.

Conan Doyle A. Sherlock Holmes, *The Sign of Four*

Hawkins, G., 'From Euclid to Ptolemy in English Crop Circles', *Bull. Am. Astronm. Soc.*, 29, p 1263, 1997

Kollerstrom N., Crop Circles: The Hidden Form, *Wessex Books*, 2002

Levengood W.C, Anatomical anomalies in crop formation plants, *Physiol. Plant* 92, 356, 1994

Levengood W.C., Semi-Molten Meteoric Iron Associated with a Crop Formation, *J.Sci.Exploration*, Vol.9, No.2.pp 191-199, 1995

Levengood W.C & Talbot N.P., Dispersion of energies in worldwide crop formations, *Physiol. Plant 105, pp 615-624, 1999.*

Meaden G.T. Circles from the Sky, *Souvenir Press*, 1991

Pringle L. Crop Circles, *Jarrold Publishing* 2004

Thomas A., Vital Signs: A Complete Guide to Crop Circle Phenomenon and why it is Not a Hoax, *North Atlantic Books* 2002

Chapter 16

Mind at Large

*Great spirits have always encountered violent opposition
from mediocre minds.*

Albert Einstein

In 'Doors of Perception', Aldous Huxley said: *"...the
brain does not produce mind, it reduces mind...each of us is po-
tentially 'Mind at Large'. But in so far as we are animals, our
business is at all costs to survive. To make biological survival
possible, Mind at Large has to be funneled through the reduc-
ing valve of the brain and nervous system. What comes out at
the other end is a measly trickle of the kind of consciousness
which will help us stay alive on the surface of this particular
planet... The various 'other worlds' with which human beings
erratically make contact are so many elements in the totality of
the awareness belonging to Mind at Large'"*

Materialistic science would have us believe that mind
is a product of the brain. However, a wide spectrum of
human experience suggests that mind and consciousness
are not confined to the brain but are more universal in nature. Aldous Huxley was referring to experiences he and
others had in the early days of experimenting with psy-
chedelic drugs. Many people have psychic experiences
without drugs which support Huxley's concept of the
'Mind at Large'. Others realise that the mind is more than
the brain when they go through a near death experience.
Documented reports of near death underpin the idea that

the mind does not rise from biochemistry in the brain but is something that expresses through cerebral neurophysiology during life and separates away at death.

Scientists and sceptics find ways to explain away the experiences thousands of people have in near death episodes but clever people in every sphere of life find ways to dismiss things they don't believe in. Disapproval is not disproof.

Dr. Raymond Moody is famous for his documenting experiences following near death. In his book 'Life after Life' he described a typical near death experience: *"A man is dying and as he reaches the point of greatest physical distress, he hears himself pronounced dead by his doctor. He begins to hear an uncomfortable noise, a loud ringing or buzzing, and at the same time feels himself moving very rapidly through a long dark tunnel. After this, he suddenly finds himself outside his own physical body, but still in the immediate physical environment, and he sees his own body from a distance, as though he is a spectator."*

People who have been near to death in the operating theatre or in a motorcar accident report being out of their bodies, fully conscious, looking down on the scene of the operation or the accident: *"I saw them resuscitating me. It was really strange. I wasn't very high; it was almost like I was on a pedestal, but not above them to any great extent, just maybe looking over them. I tried talking to them but nobody could hear me, nobody would listen to me..."*

Of particular interest are the reports that patients were aware of what people were saying and thinking about them: *"I could see people all around, and I could understand what they were saying. I didn't hear them audibly, like I'm hearing you. It was more like knowing what they were thinking, but only in my mind not in their actual vocabulary. I would catch it the second before they opened their mouth to speak."*

If these reports are accepted it is clear that the mind can exist independent of the brain, that it is still aware of the physical world and perceives thoughts of others more acutely than when confined to the brain. This supports the Huxley view that the brain is a reduction filter of the mind rather than the source of the mind.

Near death experiences support the idea that the human mind is a manifestation of super-energy in a greater reality that overshadows the brain according to the law of simultaneous existence. When the brain dies the cognitive mind then returns from containment in the human body to the universe at large: *"...There was a lot of action going on, and people running around the ambulance. And whenever I would look at a person to wonder what they are thinking, it was like a zoom-up, exactly like through a zoom lens, and I was there. But it seemed that part of me – I'll call it my mind – was still where I had been, several yards away from my body..."*

From within the super-energy continuum of space and time it would seem the mind can re-enter the subset physical space and time at any point: *"When I wanted to see someone at a distance, it seemed like part of me, kind of like a tracer, would go out to that person. And it seemed to me at the time that if something happened any place in the world that I could be just there..."*

According to the law of subsets, the form of super-energy, we call mind, could resonate with the brain, much as a broadcast programme resonates with a radio or television set. A primitive person watching a television or listening to a radio for the first time might assume the voice and pictures are products of the sets. Someone more knowledgeable would explain that the programmes are carried by invisible waves of energy from elsewhere that overshadow the radio or television set and animate them.

The materialistic theories of mind as a product of the brain are tantamount to the primitive assumptions of people who don't understand the way televisions and radios work because they cannot see or touch the invisible broadcast signals.

Broadcast programs can be received by more than one radio or television set. In like manner it should be possible for more than one brain to resonate the same quantum 'packet' of mind we describe as the human personality. If the personality departs from one body at death it is theoretically possible for it to re-appear in another at birth.

The laws of simultaneous existence and subsets allows for the reincarnation of personalities. Outside physical space time the mental personality could look into our subset world and choose to overshadow and resonate through different bodies in our past, present or future.

The same personality could choose to resonate with and animate a human body in different times and different cultures to set up 'consecutive incarnations.' The same personality could also 'incarnate' through a number of different bodies in the same time and culture but in different places to set up 'simultaneous incarnations.'

You, as a person, could inhabit bodies in ancient Egypt, Rome and Native America as well as the body you live in today. However, you could also be incarnate today in a body in England and another in South America. Because of amnesia you would not be aware of the simultaneous existence of your personality in two bodies. However, for your 'mind at large' it may be important to for you to experience being in more than one culture in these unique, transformative times. The Earth is carrying the greatest population in its history. This facilitates the greatest opportunity for simultaneous incarnation. You could fall in love with someone unaware that they are you in another body or you might assault yourself in another body.

The paradox of 'reincarnating to shoot your father or mother' is resolved by amnesia. Part of the human condition is the state of not knowing who we really are or where we come from. In the progress of childhood comes forgetfulness of other worlds. In reality every human incarnates the same 'one consciousness' so ultimately we are all the same person in billions of different bodies. In this context all theories and debates about reincarnation are irrelevant. What matters is that incarnation creates individualisation and reincarnation allows for possible continuity of individuals for growth and evolution through many different experiences in different times and cultures. Maybe one lifetime is insufficient for all the lessons and balancing necessary for an individualised stream of consciousness.

A good analogy for reincarnation is progressing from one car to another. The conscious personality would correspond to the driver and the body to the car. Birth would be equivalent to buying a new car, life to driving it around and death to scrapping it. Reincarnation would be the equivalent to replacing the old model with a new one.

Practically all world religions teach of the continuity of human consciousness after death in other worlds and many teach of reincarnation into this world. Though many people disbelieve in reincarnation evidence has been gathered in support of the theory.

Dr. Ian Stevenson of the University of Virginia carried

out extensive case studies of reincarnation. Over a period of twenty years he collected over 2,000 cases in support reincarnation and from 1960 onwards he published his findings in over twenty books and in numerous journals. Stevenson emphasises that his findings do not prove re-incarnation but provide a body of evidence supportive of belief in the idea.

Dr. Stevenson concentrated his researches on the testimony of small children. He chose cases where children had spontaneous recollection of a past life that revealed details that could be cross checked.

One case study was a small boy in France born with a number of small birthmarks. As soon as the child could speak he claimed the marks were left by bullets that killed him. As his speech developed he named the men who accused him of cheating at cards then shot him. He detailed members of his family, the name of his girl friend and the village where he lived in Sri Lanka. His French parents had a difficult child who insisted on eating with his fingers and demanding rice and curry. He wrapped a cloth round himself fastening it like a sarong, broke into Sinhalese and climbed trees with agility in search of coconuts. Subsequent investigation revealed that a coconut picker had been shot for cheating at cards, in the Sri Lanka village named by the boy, a few years before his birth in France. At the age of five the memories faded and the child grew up normally in the French household.

Dr. Stevenson – and a Dr Karl Muller, who also investigated cases where children spontaneously recalled lives — said it is not uncommon for children between the ages of two and four to speak as though they have had a previous existence. The problem with these cases is children of that age find it hard to express themselves or be taken seriously by their parents and by the time they are articulate the memories fade.

Another researcher Dr. Frederick Lenz investigated cases of people who claimed to remember a period of existence in a non-physical world. These people recalled death in a past life, passage through other worlds after death and subsequent rebirth. Lenz found the descriptions fitted closely to accounts in the Tibetan Book of the Dead.

The report of one person read: "*I felt that all my life I had been dressed in a costume but I didn't know it. One day the costume fell away and I saw what I really had been all along. I was not what I thought I was. All my life I had thought of myself as a body and a person...When I woke up in this world I realised I was not those things...It was like waking up from amnesia. I was overjoyed to be 'me' again. I had been all along but I had lost sight of it and thought I was the physical body. My body was only a thing I used for my life on earth. When it wore out I got rid of it.*"

Many people have experiences like this and identify the physical body as a prison that traps the real 'me' for the duration of life on earth. People with out-of-body, near-death or other-world experiences identify themselves more as a bubble of consciousness, a mind at large than as a physical body. As consciousness breaks free from the body many such people experience it as though waking from a dream: "*I found myself in a vast place. I felt as though I had come home. I had no apprehensions, fears or worries. I no longer remembered my former life on earth. Nothing existed for me but quiet fulfillment. I was not conscious of time in the usual sense; everything seemed timeless. I felt as if I had always been there. It was similar to the feeling I have when I wake from a dream that seemed very real only to realise it wasn't real, it was only a dream. That is how I felt. My former life on earth had been a passing dream which I had now wakened from.*

People experience life beyond the body as dwelling in consciousness. They record passage between levels which fits with the prediction of an ascending series of quantum realities in the Universe: *"I did not have the sense of moving through space. Everything was consciousness and pure awareness... I moved through thousands of levels. On each level different souls were resting before being born again. The lower levels were much darker. I somehow knew that the souls on these levels were not as mature as those on the higher levels. Finally I reached a level that I was comfortable on. I stayed there. I sensed there were many levels above the one I stopped at and that souls more advanced than I would go there..."*

These reports are too subjective to be taken as evidence of a 'soul' that survives death. However there are objective studies of a 'field' associated with the body that appear to correspond closely to the traditional concept of a soul.

Lenz, Frederick, *Lifetimes,* Bobbs-Merrill N.Y., 1979

Moody, Raymond, *Life after Life,* Bantam Books (1967)

Stevenson, Ian, *Cases of the Reincarnation Type and Twenty Cases of Suggestive Reincarnation* University of Virginia Press

Chapter 17

Fields of Life

True religion is real living; living with all one's soul, with all one's goodness and righteousness.

Albert Einstein

Harold Saxton Burr, emeritus professor of anatomy at Yale University, measured fields in and around living organisms, which he called the 'Life Field' or the L-field. In his extensive research he determined that the field had a major impact on the process of differentiation or morphogenesis. At conception a single cell contains the genetic information for every single future cell in the future multi-cellular organism. The question is what determines one cell becoming a brain cell and another, a cell in the skin.

There are many theories but Burr's extensive research revealed that the process of differentiation is influenced by a field associated with the organism from the moment of its conception. In Burr's own words, *"When a cook looks at a jelly mould she knows the shape of the jelly she will turn out of it. In much the same way, inspection with instruments of an L-field in its initial stage can reveal the future 'shape' or arrangement of the materials it will mould. When the L-field in a frog's egg, for instance, is examined electrically it is possible to show the future location of the frog's nervous system because the frog's L-field is a matrix which will determine the form which will develop from the egg."*

In the words of Dr. Northrop who worked with Burr, *"The growth and development of an embryo would seem to be the result of the fact that some kind of a factor sits on top of the embryo during its entire development and gives it direction."*

According to Burr, the location of a cell in the L-field is as important as the genetic information it contains. The L-field is like an energy blueprint which directs each cell in its development into a specific type of tissue.

In the L-field, Burr believed that he had solved the mysteries of biological life. Most scientists, however, have rejected this conclusion. Burr's interpretation of his results is highly controversial because it implies that there is more to life than mere biochemistry. It calls into question the most cherished doctrines of materialism as it implies the existence of independent energy field directing the development of living organisms.

Burr's discoveries had been anticipated by 'vitalist' philosophers in the 1920's who coined the term *élan vital* to describe the animating principle behind life. During the 1960's Russian scientists developed a technique called *Kirlian photography* for taking vivid colour photographs of the energy patterns surrounding living organisms. These patterns have been shown to persist around amputated limbs and leaves.

Rupert Sheldrake is famous for his prediction of a *morphogenetic* field responsible for the development of tissues from cells. Sheldrake is able to answer many questions in biology for which there are no answers in the orthodox science. Sheldrake takes his concept of morphic fields beyond space and time. He envisages an interlocking hierarchy of interlocking fields extending from atoms through cells and organisms to blueprints for entire species. Sheldrake's ideas fit closely with the concept of super-energy occurring as a series of quantum realities, each a subset of the one above it in an ascending series.

Dr. Masaru Emoto of Japan has made an extensive study of ice crystals and has shown that the formations are very different in different waters from different situations. His research reveals graphically the memory of water, how it seems to know where it has come from and how it has been treated.

Masaru Emoto has shown that water when frozen, thawed and refrozen crystalises in the same pattern. This confirms that something is controlling the differentiation of non-living as well as living systems. Emoto's research goes a long way to confirm Shledrakes theory of the morphogenetic field.

Emoto's studies reveal that ice crystals form in different characteristic ways in response to human thought and intent. This amazing discovery suggests that the hydrogen bonds in water that influence the way water crystalises are capable of resonance with thought. The hypothesis of super-energy fields overshadowing and resonating with water molecules to influence their formation as ice would account for Emoto's findings and his research, in turn, would appear to support the laws of simultaneous existence and subsets. Emoto's studies show a strong link between the super-energy field and the mind.

The 'Fields of Life' could be fields of super-energy coinciding with a physical body according to the law of simultaneous existence. Information could then pass from the super-energy field into the body according to the law of subsets to influence the way cells differentiate and grow. The question is, 'how would this occur? What would be the mediating factor between the field and the cell through which the information could pass and act as an instruction to the cell?

Fritz-Albert Popp, assistant professor at Marburg University in Germany, discovered that chemicals and living organisms emit and absorb photons of light and that these bio-emissions had an influence over health and disease. His work has helped provide ways of understanding why organisms are influenced by certain frequencies of light and how photons are generated and absorbed by plants and animals. Popp's 'biophotons' are resonance phenomena and his research suggests that DNA has 'resonance capacity'.

We know that differentiation involves switching on genes and complex chemical messages. However, these are more the consequence than the cause of the process. The underlying direction of differentiation is still a mystery. What is the orchestrating factor? How would it be linked to DNA and the process of resonance?

In 1913 Dr. Abrams applied the principles of oscillating electrical circuits to biological organisms through a system of diagnosis and healing called 'radionics'. The central element in a radionics box is an oscillating circuit; the basis of a radio set e.g. an electric coil and variable capacitor, which tunes the coil for resonance. Radionics circuits make no sense electrically because they are isolated. However the do make sense in terms of resonance with fields of super-energy. The clue to their understanding lies in the structure of DNA.

The DNA molecule is a double helix coiled upon itself several times. This structure is very reminiscent of a radio coil, the key element in resonant electrical circuits. The DNA molecule isolated in each cell is akin to the coil isolated in a radionics circuit. It is possible that the DNA molecule is resonating with an invisible field of super-energy in much the same way that a tuned radio coil will resonate with the invisible electro-magnetic fields of radio waves.

The idea of DNA resonance is speculative but it points to a new direction of research in biology that could throw considerable light on the process of morphogenesis or differentiation of living organisms that could revolutionise medicine; especially the prevention and treatment of cancer.

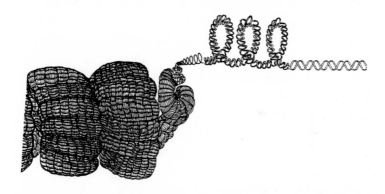

It could be that information is passed from a super-energy life-field to the DNA molecule by resonance in much the same way that a program is passed from a broadcast carrier wave to a tuned coil in a radio or television set. In its chemistry the DNA molecule carries the codes for differentiation. In its physical structure it could resonate with fields of super-energy enabling information that influences the activation of these codes. This could direct the operation of DNA causing the cell to differentiate in a certain way. Resonance could also be occurring in other organelles such as the mitochondria where strands of DNA and RNA occur.

Peter Hewitt, co-author of '*The Vortex: Key to Future Science*', came up with the idea of a 'life-program'. His concept was that a super-energy field overlaying the cell has a specific program that instructs the cell, via DNA resonance, in its structure and function. The set of frequencies for each cell then set up a field-of-frequencies resonating with the body as a whole. If this idea is correct, the physical body would be a physical manifestation cell-by-cell, of a 'frequency body' in super-energy. Any disruption on the life-program would pre-dispose the body to disease and termination of the program would result in death.

DNA resonance could also be the mediating factor in 'intelligent evolution'. Just as information is projected from the realm of super-energy to cause patterns in corn fields so information could be broadcast to DNA molecules of a particular species causing a new genetic sequence to be added to the DNA, instructing a change in the species. Resonance would explain how such an evolutionary change could occur simultaneously in a large number of species individuals. Televisions tuned to the same channel respond to the same change of program on the broadcast signal. So cells tuned to the same species-specific channel of super-energy would respond to any 'design change' in a broadcast program.

The appearance of chance in the process of evolution could be accounted for by DNA resonance. Imagine for a moment that there were intelligent beings in the realms of super-energy making design changes to DNA through DNA resonance. How would these changes appear to us? According to the law of subsets, beings operating through DNA resonance would be able to see us whilst we could not see them therefore their interventions in our world would appear as accidents; pure blind chance.

Peter Hewitt used the analogy of 'chickens in the garden'. Imagine chickens confined in an enclosure by a wire fence. The enclosure is part of the garden. You can go into the enclosure but the chickens can't get out into the garden or roost in your bed so whereas the enclosure is part of your continuum, the garden and house outside the enclosure would not be part of the chicken's space-time continuum. You give them feed and water and take their eggs. Periodically, when you need meat, midst squawking and a flurry of feathers, one from amongst them disappears never to be seen again. From the chicken's 'bird brain' perspective, these events are a mystery. They are chance events. Food and water just happen to arrive. The eggs disappear and the issue of disappearing chickens is a matter of pure blind chance for the unfortunate roosters. From your greater perspective none of these events would be happenchance. The feeding and watering of the chickens, the collection of the eggs and slaughter of roosters would be carefully planned and well orchestrated.

DNA resonance provides an explanation for the sudden appearance of new species, which cannot be accounted for in Darwin's theory. Peter depicted this allegorically in the mythology of Pan. The pipes of Pan symbolised for him resonance and the tune symbolised the super-energy 'life-program' directing the genetic process. He said, *"When Pan plays the same tune, the species remains the same. When Pan plays a different tune a new species emerges"*.

The model of a super-energy field, animating a body by DNA resonance, fits the image of a soul. In tradition the soul was the person and the primary vehicle of consciousness animating the physical body. The soul was believed to survive death to live in another world and it was the soul that carried the spirit, the spark of divinity. The soul would correspond to a field of super-energy resonating with the body via DNA and the spirit to the animating consciousness.

The law of simultaneous existence accounts for the coincidence of the super-energy 'soul' with the physical-energy 'body' and the law of subsets supports the idea of the physical body being a part of or lower aspect of the super physical soul.

The law of subsets also suggests that we are 'spiritual' beings having a physical experience more than physical beings having a spiritual experience. From that perspective, everything that happens is perfect because intelligence in super-energy benefits from a full spectrum of experiences in physical reality, all of which are relevant.

Ash D. & Hewitt P., *The Vortex: Key to Future Science*, Gateway Books 1990

Burr H.S. *Blueprint for Immortality*, Neville Spearman, 1972

Davidson J, *The Web of Life*, C. W. Daniel 1988

Emoto M. Messages From Water Vols I & II, HADO Tokyo, 1999

McTaggart, Lynne, *The Field*, Harper & Collins 2001

Popp F.A. Biophoton Emission, *J. Photochem & Photobiol*, 1997; 40: 187

Sheldrake Rupert, *A New Science of Life*, Paladin Books, (1987)

Chapter 18

Alternative Medicine

A scientific truth does not triumph by convincing its opponents and making them see the light, but rather because its opponents eventually die and a new generation grows up that is familiar with it.

Max Planck

DNA resonance could provide an obvious and simple account for the way alternative medicine works. Most alternative medical practices recognise the existence of subtle energy fields that overlay the body and influence the processes of disease and healing.

Disease is basically a breakdown in the systems and organisation of a multi-cellular body. Healing is the process of reorganization of cells so that they perform vital functions at optimum level. All these processes are influenced, ultimately by DNA. Direction by the super-energy field, via DNA resonance could have a major effect in stimulating the self-healing of a body.

The concept of DNA resonance could be valuable in cancer research. Cancer is a breakdown in the process of differentiation as undifferentiated cells multiply into tumours. This disease could be caused, in part, by weakness or breakdown in the life program caused by negative thoughts, stress, electro-magnetic pollution and the environment in which people live. This would impact on the maintenance of differentiation by DNA resonance.

DNA resonance could be considered in the prevention and treatment of cancer alongside the danger of carcinogenic chemicals, genetic predisposition, microbes, malnutrition and free radicals that impact cells directly. As medicine evolves it will take into account energy influences on the cell as well as chemical.

DNA resonance could provide a rationale for healing. Every age and culture has recognised the existence of subtle energies that overlay the body and influence the process of natural healing and tissue regeneration. These include orgone energy described by Dr. Wilhelm Reich, the morphogenetic field of Rupert Sheldrake, the life-fields described by Harold Saxton Burr. This would correspond to the healing energy spoken of by people such as Mary Coddington. Native people recognised these subtle energies calling them the mana of the Polynesian Hunas and the orenda of the Native American Iroquois.

Super-energy fields, in and around the body, have been seen as an aura by psychics, and described as Prana in India. This life force is ruach in Hebrew, barraka in Islam, chi in China and ki in Japan. Paracelsus called it archaeus and Mesmer called it animal magnetism.

Mesmer's description of the super-energy field as 'animal magnetism' is suggestive of symmetry between fields of physical and super physical-energy. If the principle of symmetry were applied to fields of super-energy they would be expected to behave in much the same way as electric and magnetic fields. For many thousands of years Yogis in India have related the healing to magnetism and magnets are considered by many alternative practitioners to have healing properties.

From electro-static research it is clear that electric fields are evenly distributed over spheres. However, if there is a spike on the sphere, the charge will accumulate on the spike and a potential difference will occur between the spike and the sphere. Symmetry suggests that these electro-static principles would operate in the super-energy fields.

As a super-energy field overshadows the contours of the physical body electro-static principles would suggest that super-energy potential gradients would occur between the main body of head and torso and the appendages such the fingers and toes, hands, feet and ears, occurring as 'spikes' on the body.

Energy, in the form of electricity, flows down a potential gradient. Super-energy would be expected to flow down the potential gradients between the body and peripheral points where charge in the super-energy field would be expected to accumulate. Professor Harold Saxton Burr measured potential gradients between the digits, limbs and the body.

The Chinese discovered life-energy, or chi flow lines, associated with every organ in the body. These lines running from the head and limbs to the body through the skin they called meridians. In Chinese medicine the meridians are likened to rivers and the organs to lakes. Just as a lake is influenced by the flow of water in the river so the organ is affected by the flow of chi – vital energy – in the meridians. They found that by stimulating or sedating the flow of chi in the meridians they could restore energetic balance in the organs. The insertion of needles for this purpose, into specific points on the skin, related to the meridians, evolved into the Chinese system of medicine called Acupuncture.

In Acupressure, the flow of chi is induced by massaging specific acupressure points on the meridians. As the organ-specific meridians terminate on the soles of the feet, acupressure on the terminal points would be another approach to this therapy. Such a therapy exists. It is called Reflexology.

In 1986, the British Medical Association published a report on alternative medicine. Whilst they could not deny that alternative therapies were effective, they could not support them because they were considered to be unscientific. The BMA had especial difficulty in accommodating reflexology. They said that they could find no rational basis for it in medical science.

Alternative medicine works. Reflexology works. People would not pay for alternative therapies in competition with the free medical service in the UK if they didn't. The problem for the BMA is that medical science is stuck with the outmoded paradigm of materialism.

To understand alternative medicine we need to embrace a completely new scientific paradigm based on a true understanding of 'multi-dimension' energy. We need to think 'outside of the box' of physical energy and allow for the influence of energies existing in 'higher dimensions' or higher quantum states of reality.

With the super-energy model it is easy to accommodate healing. Healing would be the direct transfer of super-energy from the field of the healer to the field of the patient. Distant healing is possible because the super-energy fields exist outside of physical space and time and so the terrestrial separation between the healer and the patient has no bearing on the field interactions. Intent by the practitioner has more significance than spatial separation in healing. This is to be expected if the underlying nature of subtle 'super-energy' is mind.

In 'Reiki' sacred symbols are influential in the healing modality. In his research Masaru Emoto discovered that symbols affected the way water crystalises in much the same way as thought and intent. It would seem that patterns in the physical have an impact on patterns in the super-energy field. This is apparent in homeopathy.

Homeopathy was founded 180 years ago by a German, Dr. Hahnemann on the principle of *'like cures like'*. Hahnemann found that if a substance, that caused a particular set of symptoms, was greatly diluted the diluted form of the substance would cure the symptoms. For example, arsenic will cause severe stomach pains but Hahnemann realised that a dilution of arsenic would cure stomach pain. Hahnemann discovered that the more he diluted his remedies, the more effective they became if he potentised them; on each dilution he potentised his remedies by percussing them. In some way he found the agitation increased their potency.

Homeopathy could operate on the principle that 'anti-like cures like'. The potentising process could act in a similar way to photography. In the process of potentising, the super-energy field of the substance, that causes the symptoms, is impressed on the super-energy field of the substrate, in which it is diluted. The field pattern of the substance acts as a positive and in the potentising process it creates a super-energy negative. With increased dilution and potentisation, the positive is decreased and the negative pattern increased. The concentration of symptom causing substance is reduced whilst the quantity of anti-symptom substrate is increased. This is why homeopathic remedies become more effective as they are diluted and potentised.

Homeopathic medicines act energetically rather than chemically. This was clear from the research of Professor

Jacques Benveniste. In 1984 Professor Jacques Benveniste, quite by accident, proved in the laboratory the basic principle of homeopathy that a substance can still have a therapeutic effect even when diluted to the extent that not a single molecule of the original substance remains.

Benveniste demonstrated that water has a memory and can take on properties of an original substance that seemed to be amplified by dilution – the central principle of homeopathy. He went on to reveal that molecules each have their own frequency of vibration and will resonate with other molecules of similar wavelength. Benveniste discovered that water molecules were resonating with the frequencies in the audible range. Substrates like water appeared to have the ability to pick up the characteristic frequency of a substance and in the process of dilution, the frequency in the substrate increased. Benveniste's work was reinforcing Popp's biophoton discoveries and fitted with the premise of radionics and homeopathy that substances have specific frequencies, which can be used in place of the substances in the diagnosis and treatment of disease. All this configures with the hypothesis of super-energy fields resonating with physical matter.

Most alternative medical therapies make sense when viewed as a way of treating the physical body through the 'super-energy' life-field. The field finds its own level and influences differentiation wherever there is imbalance and a need to stimulate cellular regeneration. The ability of the practitioner to transmit the subtle energy is more important than the particular form of therapy practiced. Ultimately, in all forms of healing, it is the state of being of the conduit that counts.

When she heard about Dr. Emoto's work, Anna –

mother of six of my children — sent me an email:

"Now, you know that every snowflake is unique? But did you also know that if you take one snowflake crystal and melt it, then freeze it again, it will reform exactly as it was before it melted! That blows my mind. There is such an order in this intricate vast mystery, a sort of preordained pattern in everything. It's like we all have our destiny and our potential for vibrant health, so it doesn't matter what comes along to change it or try to destroy it, if we have faith in the infinite re-creative energy real healing can happen."

The shadow to DNA resonance is electromagnetic smog. The electro-magnetic radiation of wifi transmitters' and mobile phone masts could be hazardous to health. Increasingly research is revealing this to be the case.

Davidson J, *Subtle Energy*, C. W. Daniel 1987

McTaggart, Lynne, *The Field*, Harper & Collins 2001

Sheldrake Rupert, *A New Science of Life*, Paladin Books, (1987)

Chapter 19

Ascension

Men occasionally stumble over the truth, but most of them pick themselves up and hurry on as if nothing had happened

Sir Winston Churchill

If the universe is just like the atom, because electrons in the atom can take a quantum leap from one level to another it should be possible for bodies of matter on one energy level of the Universe to make a quantum leap to another.

The difference between the levels of energy in the Universe is a factor of speed not frequency. Each plane or level would possess a full spectrum of frequencies in light and sound. The difference between physical-energy and super-energy on the planes is the intrinsic speed of energy.

In our world the intrinsic speed of energy is the Einstein constant — the speed of light. This is why space, time, and mass are relative to the speed of light.

If bodies of matter were to make a quantum leap from one plane to another in the Universe it would be by change in the intrinsic speed of energy in every vortex and wave of their atoms. This process can be called the *ascension* as it causes bodies to ascend from physical-energy to super-energy in the ascending series of quantum realities.

The ascension process would occur by increase in the intrinsic speed of energy in the atom, not by a change in frequency of vibration. Frequency change is merely a metaphor for the process invented by Sir Oliver Lodge in the 19th century. What evidence, if any, is there of the ascension process?

A controversial area of alternative medicine is 'psychic surgery'. After initial reports of the phenomenon in the media the entire subject was discredited by a spate of 'fraudulent psychic surgeons'. This was similar to the pattern of disinformation used to discredit crop circles; Initial reports quickly followed by hoaxes. Hoaxing creates confusion. People don't know what to believe. Hoaxing is employed when awkward facts threaten the scientific and religious establishment. The public believe reports in the media and the opinion of experts. Undiscerning people believe things that are discredited so it is easy to manipulate public opinion by setting up a hoax as soon as a paranormal phenomenon is reported. This destroys credibility in the public eye and maintains public support for scientific materialism and established theology.

It is very difficult to substantiate predictions pertaining to super-energy when evidence is clouded with suspicion and disbelief. The authenticity of crop formations has been established by sound research. Because psychic surgery depends on a human operator it is more difficult to establish impartial, objective proof for its validity. However, for those who are satisfied with genuine operators, psychic surgery belongs to a class of phenomena that supports the ascension prediction that if the intrinsic speed of energy, in every vortex and wave in matter is accelerated beyond the speed of light then the matter takes a quantum leap into another level of reality.

In psychic surgery a wound is reported to appear by means of which a surgical operation is performed. At the end of the operation the wound heals instantly with no bleeding and no sign of a scar.

The ascension account for psychic surgery is that the speed of energy in the vortices and waves of each and every atom and molecule in a selected area of skin and muscle of the patient goes beyond the speed of light. The energy in this matter is ascended into super-energy. Physical light is not reflected off super-physical matter so the tissues become invisible. No longer interacting with the force fields of physical matter they also become intangible. The selected area of skin and bone, muscles blood vessels and nerves vanishes out of physical space and time leaving an apparent void.

The psychic surgeon is able to take advantage of the sudden appearance of a hole in the patient's body to perform an operation — such as the removal of a tumour. In this extraordinary process, the tissues do not go anywhere. They merely pass beyond human perception. In the course of apparent dematerialisation there has been no change in the atomic or molecular structure of the cells. They have become a 'micro-zone of super-energy'. Physical blood flowing through the vessels into the zone leaves space and time to become 'super-physical blood'. On leaving the zone the blood corpuscles and plasma revert to physical blood. With no real wound, and no break in the blood vessels, blood continues to flow through the super-energy space-time zone unaffected by the change in the intrinsic speed of its energy.

If there were change in frequency in the process electrons would undergo quantum leaps in the atoms which would cause disadvantageous chemical reactions such as coagulation of proteins. In simple language increase frequency of vibration in matter causes temperatures to rise and tissues to cook.

On completion of the operation the psychic influence causing energy to ascend to super-energy is removed. Reverting to physical matter the tissues reappear in physical space and time. Instantly the apparent wound heals. In fact there was never a wound in the first place and therefore no healing was necessary. All that happened was that some skin and bone temporarily ascended in the universal quantum series causing it to vanish. Reversal of the process caused it to descend the quantum series. This resulted in the tissues re-appearing again.

Psychic surgery is an advanced level of science that we have yet to fully comprehend. Refusal to entertain phenomena that impacts our limited view of reality is ignorance. The word ignorance comes from *'to ignore'*. The attitude of Robert Oppenheimer, father of the atom bomb, when confronted by David Bohm's work on implicate order was typical ignorance: *"We can't find anything wrong with it so we will just have to ignore it."*

Imagine you had a time machine and appeared in the middle ages with a mobile phone. Some people would worship you as an oracle in communication with the gods. Others would want to kill you because of the threat posed to accepted beliefs and social order. Most people would want the uncomfortable challenge to their normality removed. Relieved to hear that you have been arrested and denounced as a witch, they would gather at your pyre.

Throughout the ages men and women have appeared with mysterious powers. They have been called gods or magicians, witches, shamans, saints, sorcerers and suchlike. People are born with, or acquire paranormal powers that enable them to change physical-energy into super-energy. They can dream the world a little different. Aboriginal Australians speaking of our world as a dream-state, appreciate some people have the power to change the state of reality as though it were a dream.

If every human being is an incarnation of the consciousness that creates and controls all the energy in the Universe, it should not come as a surprise to find in history people who are aware of their innate power and use it. An outstanding example of such a person in modern times is Sri Sathya Sai Baba.

Since childhood, Sai Baba has been able to cause objects to appear and disappear at will. Scientists from around the world have visited him in his ashram in southern India and reported on his abilities.

Erlandur Haraldsson professor of psychology at the University of Iceland and Karlis Osis, Chester F. Carlson Research Fellow of the American Society of Psychic Research began their studies of Sai Baba in 1973. After materializing a rare double rudraksha bead Sai Baba transformed it into a jeweled ornament. In their own words:

"After we had admired the rudraksha, Sai Baba took it back in his hand and turning to Dr Haraldsson, said he wanted to give him a present. He enclosed the rudraksha between both his hands, blew on it, and opened his hands toward Dr. Haraldsson. In his palm we again saw a double rudraksha, but it now had a gold ornamental shield on each side of it. These shields were about an inch in diameter and held together by golden chains on both sides. On top of the shield was a golden cross with a small ruby affixed to it. Behind the cross was an opening so that this ornament could be hung on a chain and worn round the neck. A goldsmith later examined this ornament and found that it contained 22-carat gold...a botanist's microscopic examination of the rudraksha showed it to be a

genuine example of the species."

The scientists repeatedly asked Sai Baba to perform a 'miracle' under controlled scientific conditions. One day, during an interview with Sai Baba an enamel picture of Baba in a ring Dr. Osis was wearing vanished. There had been no contact between Sai Baba and the scientists before the incident and they were sitting several feet apart. The fastenings in the ring were intact and though they searched the room no picture was found. Sai Baba said this was his little experiment. He asked if Dr. Osis wanted the picture back or a different picture. Dr. Osis asked for the same picture. Sai Baba blew on the ring. The original picture returned but the ring was different.

On their return to America, Osis and Haraldsson consulted the celebrated magician Douglas Henning about this incident. He said most magicians could manifest objects by sleight of hand but producing things by demand required extraordinary skill. However, the original picture appearing in a ring totally transformed around it was beyond any magician's tricks he was aware of.

Sai Baba says he visualises the objects he manifests. Until then they are stored in his mind in what he refers to as the 'Sai stores'. This fits with the description of the Universe as a mind. Objects in super-physical reality exist as memories. These can manifest as physical reality if the intrinsic speed of energy in the 'thought forms' descends to the speed of light. In another level of quantum reality these memories exist as atomic matter and life forms. As such they can appear and disappear in our world merely by change in intrinsic speed of their energy.

The power to change the intrinsic speed of energy is innate in us all as we are all manifestations of the consciousness that imagines energy into existence. In religious tradition the power to change the speed of energy and work apparent miracles was called 'grace'.

Visualising food enables Sai Baba to feed thousands out of a small pot. The only miracle Amma performed at the beginning of her ministry involved feeding more than a thousand people pudding from a small brass pot. If the Gospel stories are to be believed, Jesus fed five thousand people on a few loaves and fishes. He might have used the same power of 'mind-over-matter' employed by Sai Baba and Amma today.

If grace enabled mystics to cause objects to appear and disappear, perhaps the same could happen with their own bodies? If the speed of energy in the atoms of a mystic's body were to accelerate beyond the speed of light the mystic would vanish from our world and appear on another plane of quantum reality. Leaving physical time as well as space they would step out of the time stream that controls their aging.

In France, before the revolution, the mysterious Count St. Germaine was reported to have had the ability to manifest diamonds for destitute nobles. He would also vanish in one place and reappear elsewhere and he never seemed to age. Throughout the period 1723 - 1789 when he was active in France, he always appeared to be aged 45 to 50. His death was recorded on the 27th February 1784 and he was supposed to have been buried on March 2nd at Eckenforde, but he appeared at a conference, in Wilhelmsbad, in 1785 and was active again in France, in 1788, warning the nobility of the danger of revolution. He left for Sweden in 1789 and was seen again, on the eve of the murder of the Duc de Berri, in 1820. He still appeared to be about 50.

Pythagoras was reported to have appeared on the same day to groups of disciples, in towns separated by several hundred miles. History recalls his adept Apollonius of Tyana disappearing from his trial in Rome for treason. This happened in front of Caesar and is part of the Roman record. Apollonius reappeared, the same day in Puteoli, three days journey away and lived to a ripe age. In the Muslim tradition Mohammed vanished from Mecca and appeared on Mount Moriah in Jerusalem on the site now covered by the Dome of the Rock.

The common thread between Jesus and Mohammed was the celestial visitor Gabriel. Gabriel was instrumental in the alleged virgin conception of Jesus and transit of Mohammed between Mecca and Moriah. Was there a link in the two processes? Was a sperm translocated from Joseph to Mary by the same process as the body of Mohammed was translocated between Mecca and Jerusalem? Was it the same super-energy technology that enabled Gabriel himself to appear and disappear? Who was Gabriel? As an angel this celestial visitor could be described as extraterrestrial. History is littered with myths of extraterrestrial visitors. Described as gods or angels they appear and disappear and they come from the sky or the heavens. In modern language we would say they come from outer space. There is no lack of evidence of extraterrestrials but the evidence has been dismissed and suppressed by disinformation.

Scientists have dismissed the possibility of extraterrestrial visitors on the grounds that space is too vast to be traversed. With the enormous distances between stars and galaxies in the Universe, travel from one planet to another through space — intra-space travel – is a practical impossibility. Scientists do not deny the possibility that we may not be the only intelligent population in the Universe. With hundreds of billions of star systems in the average galaxy and billions of galaxies in the Universe, there could be Intelligent Populations — IPs – elsewhere in the Universe. We may qualify as moderately advanced IPs, existing on a small planet, in a modest star system, on the outskirts of a galaxy, which have just about made it to the moon by means of crude, gas emission, intra-space travel.

IPs from other star systems could be more advanced than we. They could be employing a 'super-energy technology', which enables them to reach Earth from elsewhere in the Universe. They may have mastered the technology of inter-space travel — the ability to move bodies in and out of different space-times. The possibility of super-energy technology offers scope for unlimited travel between star systems and galaxies. Inter-space travel between one space continuum and another would enable beings to move bodies up and down the ascending quantum series. This would allow intelligent populations to move themselves and their craft out of the space-time domain of one planet and descend it into the space-time of another.

A common feature of extraterrestrial UFO sightings is the way craft appear and disappear. UFO evidence, though discredited, is strongly suggestive of super-energy technology that enables craft to move in and out of space time rather than move through it. If this is true the question remains what is the technology that makes this possible?

Any intelligent population that could master the technology of ascension could transfer their bodies, in entirety and without change in structure or internal biochemistry, from physical space and time into the space-time reality of super-energy. This ascension process being perfectly safe for living organisms would offer them a viable alternative to death. Moving from one stream of time to another the ascension process could also reverse the aging process. In a single word ascension offers the promise of physical immortality. In history, beings landing on Earth from another space-time continuum would appear as immortals and be hailed as gods.

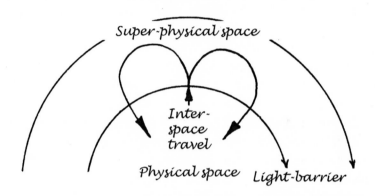

If we could master the technology of inter-space travel we would have freedom of the Universe. The question is, are we sufficiently responsible to be trusted on other planets. Consider the way we have raped and plundered the Earth with little or no regard for the environment or welfare of other species. Suppression of information may be necessary at this time. When we are wise enough to join other intelligent populations in the galaxy, disinformation will dispel. It is not for us to be concerned about suppression of information; it is for us to be concerned with our own personal growth and evolution into a species that loves and cares for itself, the planet it inhabits and other species that dwell therein. Only then will we be ready to reach for the stars. That is what ascension is all about.

Matthews, Robert, *Unraveling the Mind of God,* Virgin Books (1992)

Haraldsson, E. & Osis, K. *Appearance and disappearance of objects on the presence of Sathya Sai Baba.* J.Am.Soc.Psy.Res., Jan. 1977.

Hitching Francis, *The World Atlas of Mysteries,* Pan Books (1979).

Von Buttlar J. *The UFO Phenomenon,* Sidgwick & Jackson (1979)

Good Timothy, *Above Top Secret,* Sidgwick & Jackson (1987)

Illobrand von Ludwiger, *Best UFO Cases-Europe,* Nat. Inst. Discv. Sci.,1998.

CA Kelleher, *Retrotransposons as Engines of Human Bodily Transformation* J. Sci. Expl., 1999 V.13, No.1, p.9-24.

Chapter 20

Extraordinary Research

"I believe matter itself is just spin."

Eric Laithwaite

T ravel between the worlds by changing physical-energy into super-energy would overcome the limitations of space and time. That would open up the entire Universe to humanity. It is ironic that the technology of inter-space travel was developed by mankind during the darkest period of the 20th century.

The shifting of a naval ship, in and out of space-time, is alleged to have occurred as part of a top-secret research programme conducted in America during World War II. The project originated from a study of invisibility, in the early 1930s, at the University of Chicago, involving Nikola Tesla, the inventor of alternating current electricity, the physicist Dr. Kurtenhauer and Dr. J.Hutchinson, Dean of the University. In 1934 the project moved to Princetown for the attention of the newly formed Institute of Advanced Study, which included Albert Einstein, John Von Newmann and T. Townsend Brown.

In 1943, the US government gave the go-ahead to the infamous 'Philadelphia Experiment'. The project was intended to make ships invisible to radar and visible light. In the course of the experiment a destroyer, the USS Eldridge, vanished from the harbour of Philadelphia for 15 minutes and was reported to have reappeared in the harbour of Norfolk Virginia, several hundred miles away. The experiment was disastrous for the crew on board. Many of the sailors went out of their minds and five of them were reconfigured with the atomic structure of the ship. Their hands were merged into the metal where they had been touching it at the time of the experiment. Release was only possible through amputation. Needless to say the project was abandoned.

After World War II the invisibility project was renewed under the direction of Dr. Von Newmann, at the Brookhaven National Laboratory until Congress disbanded it in 1967. In *The Montauk Project*, Preston Nichols claims that it was re-established, in 1971, as a secret military project under the continued direction of Von Newmann at the Montauk Air force Base on Long Island.

In this extraordinary book, Nichols claimed that vortex resonance teleportation beams were constructed at Montauk, which enabled bodies to be moved in and out of time as well as space. He claimed that an inter-dimensional link between the Philadelphia experiment in 1943 and the Montauk experiment in 1983 caused parts of the ship and two crewmembers to be teleported from the SS Eldridge in 1943 to the Montauk base in 1983. This story was the central theme of a popular movie.

The Philadelphia experiment and Montauk project suggest that a 'vortex resonance' technology for the ascension of physical bodies is within the grasp of mankind. Should this involve change in the Einstein relativity constant from physical-energy to super-energy, without change in frequency of vibration, the implications are staggering.

We are all too aware of the outcome of the genius of Albert Einstein. However, other members of the Institute of Advanced Studies also made discoveries with far reaching implications for mankind.

T. Townsend Brown pioneered the anti-gravity properties of electric charge in the 1920s. He discovered that if a disc is charged so the upper side is positive and the lower is negative when mounted horizontally the disc thrusts upwards toward the positive pole, effectively acting against gravity.

1 Because gravity arises from matter antimatter annihilation it has the potential to release more energy than the sun or hydrogen bomb. This is because annihilation releases around two hundred times the energy of nuclear fusion.
2 Because gravity arises from electric attraction it should be possible to generate an anti-gravity force from electric repulsion.
3 Because gravity originates from vortex interactions it should be possible to generate an anti-gravity force from spin.

In 1929 Brown published a paper on his anti-gravity discovery that a 1% weight loss could be generated by a 100Kvolt electric field. He went on to invent a revolutionary anti-gravity motor with no moving parts. His anti-gravity disc was segmented so that each segment could be selectively charged. Moving the charge around the rim of the disc from one segment to another allowed the anti-gravity force to be directed. The implication of this discovery for aircraft was phenomenal.

John R. Searl, an electrical engineer employed by the Midland Electricity Board, investigated the anti-gravity force generated by a combination of spin and electric fields. He set a segmented rotor disc spinning through electro-magnets at its periphery. The electromagnets, energized from the rotor, were intended to boost the electro-motive force. Fortunately the generator, about three-foot in diameter, was tested in the open. The test was performed by Searl and a friend in 1952. To begin with, the apparatus produced the expected electric power but at an unexpectedly high voltage. This quickly exceeded a million volts producing a crackling sound and the smell of ozone. In Searl's own words: *"Once the machine has passed a certain threshold potential the energy output exceeded the input. From then on the energy output seemed to be virtually limitless."*

Then as the generator continued to increase in potential, it lifted off the ground and broke free of its mountings and the engine. It floated in the air, all the time spinning faster as the air around it glowed pink with ionization. Then the apparatus accelerated off into space and was never seen again! In subsequent experiments Searl mounted his generators, which he built up to thirty foot in diameter, more firmly in the ground. But they still tore themselves free of the Earth, taking the foundations with them. The hemispherical crater left in the ground, suggested the anti-gravity force was operating over a sphere with the generator at its centre. In experiments that have proved difficult to repeat, Searl appeared to have discovered a link between anti-gravity and energy. The system going out of control strongly suggests a resonance phenomenon.

Decades before Searl, the Austrian inventor Victor Schauberger, famous for his construction of logging flumes, discovered the anti-gravity properties of vortex motion quite by chance. Schauberger was a young ranger in the wilderness forest of Bernerau, in Austria, when he made his first observations of the power in vortex motion. In his own words: *"It was spawning time one early spring moonlight night. I was sitting by a waterfall waiting to catch a fish poacher. What then occurred took place so quickly that I was hardly able to comprehend. In the moonlight falling directly onto the crystal clear water, every movement of the fish, gathered in large numbers could be observed. Suddenly the trout dispersed due to the appearance of a particularly large fish, which swam up from below to confront the waterfall. It seemed as if it wished to disturb the other trout and danced in great twisting movements in the undulating water as it swam quickly to and fro. Then as suddenly the large trout disappeared in the jet of the waterfall which glistened like falling metal. I saw it fleetingly under a conical-shaped stream of water, dancing in a wild spinning movement the reason for which was not at first clear to me. It then came out of this spinning movement and floated motionlessly upwards. On reaching the lower curve of the waterfall it tumbled over and with a strong push reached behind the upper curve of the waterfall. Deep in thought I filled my pipe and as I wended my way homewards, smoked it to the end. I often subsequently saw the same sequence of play of a trout jumping a high waterfall."*

Schauberger observed that the vortex motion of water, a little above freezing, would lift the trout up the waterfall. He was also intrigued by the way trout, in the mountain streams would remain motionless, as if suspended, in the fast flowing water, then dart like lightning upstream. Schauberger was convinced that the turbulent motion of water, at its greatest density, generated a force in the opposite direction to the flow of the stream. He believed that trout could seek out the upstream flow of energy and use it to remain motionlessly suspended in the fast flow of water or to propel them upstream and over waterfalls. He believed that trout also employed a force generated by spiral motion of water passing from its gills over the surface of its body.

Victor Schauberger was convinced that the conical vortex or cycloid spiral was a source of energy. To test his idea he set himself the task of building a vortex turbine based on the same principle of twisting, reeling and spinning that he had observed in the fast flowing waters of freezing mountain streams. His most successful designs were based on the corkscrew shaped spirals expelled from the gills of trout so he called his apparatus the 'trout turbine'.

Victor Schauberger is remarkable in his ability to construct heavy apparatus without engineering training, facilities or funding. His inventions were remarkable and discoveries extraordinary.

In the early 1930's Schauberger fabricated conical

Victor Schauberger

pipes of special materials, to construct a corkscrew turbine. Operated by an electric motor, the spiral turbines screwed water into a vortex flow and directed the water onto a conventional water turbine coupled to a generator.

Schauberger discovered that the temperature of the water was critical as was the shape of his turbine and the materials out of which it was constructed. He found that as the water screwed faster it suddenly produced large amounts of energy. Coupled to a dynamo, the turbine began to generate more electricity than the input motor was consuming. The system suddenly went out of control as the apparatus tore itself away from its holdings and smashed itself against the ceiling. When Schauberger experimented with air turbines he found the same thing happened. Regardless of the medium, vortex motion seemed to generate energy, apparently out of nowhere, and it also produced an anti-gravity force.

Just before the outbreak of the World War II, Hitler took an interest in Victor Schauberger's work. He ordered a Vienna firm called *Kertl* to construct and test Schauberger's vortex turbines with a view to using them in aircraft engines. An engineer called Aloys Kokaly was employed in the manufacture of certain parts. On one occasion when he delivered the parts to the Kertl factory he was told, *"This must be prepared for Mr. Schauberger on orders from higher authority, but when it's finished, it's going out onto the street, because on an earlier test on one of these strange contraptions, it went right through the roof of the factory."*

Another inventor experimenting with free energy was the American, Joseph Newman. Newman found that free energy could be obtained by spinning electromagnetic fields. His machine consisted of a number of rotating magnets wound with copper wire to form a reciprocating magnetic armature. According to Newman, as the armature was set spinning, an electromagnetic force was induced and set into a spiral pattern of motion around the current carrying copper wire.

Like the other generators, Newman's apparatus appeared to produce energy out of nowhere. In *The Guardian* newspaper it was reported that Dr Roger Hastings, chief physicist for the *Sperry-Univac* Corporation, tested Newman's apparatus. He found that the production efficiency of the machine was far greater than 100%. On September 20th 1985 Hastings issued an affidavit to the effect that *'...On September 19th 1985 the motor was operated at 1,000 and 2,000 volts battery input, with output powers of 50 and 100 watts respectively. Input power in these tests were, 7 and 14 watts yielding efficiencies of 700% and 1,400% respectively...'*

Other inventors who claim to have developed 'over-unity' free energy generators included Bruce de Palma, Adam Trombly and Stephan Marinov. Because of the materialistic paradigm that nothing exists beyond the physical realm it is impossible to account for these discoveries in conventional science. However, if the physical is treated as part of a greater system of reality, free energy devices are not difficult to rationalise.

A clue to free energy comes from DNA resonance. DNA molecule is a three-dimensional spiral. Perhaps resonance can occur between physical-energy and super-energy through the vortex.

If the symmetry of *as above so below, as below so above* applies, and the laws of physics and forms of energy are much the same in the different quantum realities, then vortices could be prevalent in the realms of super-energy. If this is so then it should be possible to create vortices in our world, which match vortices existing beyond the light-barrier. With no space-time separation between vortices of physical-energy and super-energy, the law of simultaneous existence suggests these would coincide in the 'same here and now' so that resonance could then occur between them.

The Law of Subsets indicates that the direction of energy flow would be from the greater to the lesser. The self-evident laws of, 'water flows downhill' and 'electricity flows down potential gradients' supports the prediction that in resonance between physical and super-physical systems, the energy would flow from the super-physical system into the physical system.

Setting up a vortex for resonance would be equivalent to tuning a piano string for wave-resonance so this could be described as 'tuning the vortex'. A piano string is tuned by varying its characteristics i.e. length, tension and diameter. A vortex could be tuned by varying its speed and other characteristics such as its shape and the properties of the spinning medium.

If a vortex, with optimal physical characteristics, is set in motion and its speed is steadily increased, then at a certain threshold speed resonance could occur. Once resonance starts the energy in the vortex should increase exponentially.

The unlimited input of energy would occur because one feature of resonance is that there is no limit to the number of individual systems that can be involved in a single resonance process. Radio and television illustrate this. There is no limit to the number of radio or television sets that can resonate to a single broadcast. With no space-time separation between physical energy and super-energy, a physical vortex could resonate with every single matching super-physical vortex in existence. By doing so, the physical vortex could draw a virtually unlimited amount of energy into our world from the super-physical realms of the Universe.

Vortex resonance has the potential to provide unlimited power. At the same time, it could be one of the most destructive powers if miss-handled or misused. Look at what we have done with nuclear energy! Is it any wonder there appears to be a conspiracy, involving scientific and religious institutions, governments and the media, to suppress super-energy technologies? Everything that happens in the world is a reflection of the consciousness of mankind. Because humanity does not yet have the level of responsibility required for the quantum leap in technology that vortex resonance will bring, forces may be at work within society to frustrate progress in this direction. Even if this is so it is not for us to be concerned about conspiracies. It is for us to be concerned about ourselves. The science and the spirit of humanity can no longer be divorced. A consciousness shift is vital before we can be entrusted with super-energy.

Alexandersson Olaf, *Living Water: Viktor Schauberger and the Secrets of Natural   Energy*, Gateway Books, (1990).

Bielek A. & Steiger B. *The Philadelphia Experiment*, Inner Light.

Berlitz C. *The Philadelphia Experiment*

Cook N. The Hunt for Zero Point, Arrow 2002

Newman Joseph Westley, *The Energy Machine of Joseph Newman*, (1986)

Nichols B. P., *The Montauk Project*, Sky Books (1992).

The Biography of Thomas Townsend Brown &   Project Winterhaven The Townsend Brown Foundation

The Guardian, March 21, 1986.

Von Buttlar J. *The Philadelphia Experiment*

Wynniatt C.B., *Energy Unlimited*, Issue 20,1986.

Chapter 21

Conspiracy

Only one who devotes himself to a cause with his whole strength and soul can be a true master. For this reason mastery demands all of a person.

Albert Einstein

Thomas Townsend Brown developed an anti-gravity device early in the 20th century yet his invention never saw the light of day. Anti-gravity technology for aircraft appeared to be well advanced by the late 1950's but by 1960 all research and development in this direction disappeared from the burgeoning industries of civil and military aviation. T. T. Brown had been all but written out of the history books. In his book *The Hunt for Zero Point* Nick Cook — when talking about the B-2 stealth bomber - gives a very good reason why:

The whole science of stealth would need to be protected like no other strand of weapons science before... Eradication and disinformation, coupled with the secure compartmentalization of the programme, formed the basis of that strategy... The B-2 as an anti-gravity vehicle had been consigned to the columns of conspiracy mags and tabloids; the mainstream of aerospace press, afraid of an adverse reaction from its conservative readership, had refused to touch it.

With one exception.

Enter Britain's most eminent aerospace journalist, Bill Gunston, OBE, Fellow of the Royal Aeronautical Society and an article he had penned entitled 'Military Power'.

Gunston, who served as a pilot in the RAF from 1943 to 1948, was scrupulous over his facts – as editor of Jane's Yearbook on aero-engine propulsion he had to be. 'Military Power' published in 'Air International' magazine, was a walk-through dissertation on the development of aero-engine technology since the end of the Second World War; good Gunston Stuff, until the last couple of pages when, to the uninitiated, it appeared that the aerospace doyen had lost his mind. Gunston not only portrayed the B-2 anti-gravity drive story as fact, but went on to reveal how he had been well acquainted with the rudiments of T. T .Brown's theories for years, but had "no wish to reside in The Tower (of London), so had refrained from discussing clever aeroplanes with leading edges charged to millions of volts positive and training edges to millions of volts negative."

Gunston explained why he felt that there was much more to the B-2 than met the eye, drawing on a lifetime of specialist knowledge. In short, if you applied the laws of aerodynamics and basic math to the known specifications of the B-2, there was a glaring mismatch in its published performance figures.

It was clear, Gunston said, that the thrust of each GE engine was insufficient to lift the 376,500 pound listed gross weight of the aircraft at take-off.

The only way the B-2 could get into the air, therefore, was for it somewhere along the line to shed some of its weight. And that, of course was impossible unless you applied the heretical principles of Thomas Townsend Brown to the aircraft's design spec...

Nick Cook then continued in his own right *"During the Second World War, I remembered Thomas Townsend Brown had been involved in experiments that sought to show how you could make a ship disappear on a radar screen by pumping it with large doses of electricity. Between 1941 and 1943, Brown had supposedly been involved in tests, I saw now, that were identical in principle to the methodology that Northrop seemed to have applied to the B-2 to make it the ultimate word in stealth. Researchers had never taken Brown's wartime experiments seriously because the precise nature of the work had been obscured by the myth of the parallel dimensions aboard the USS Eldridge – the ship at the heart of the 'Philadelphia Experiment'."*

Because of B-2 secrecy there is a strong likelihood that disinformation was fed into the Philadelphia Experiment story to discredit T.T. Brown and throw serious researchers off the scent of his anti-gravity work. Fortunately, a number of other researchers discovered anti-gravity quite independently.

In October 2000, the US magazine *Popular Mechanics* carried an article on an anti-gravity device invented by an American Chinese scientist working at Alabama University. Dr. Ning Li built a disc with super-conductors. Soon after she left the university to develop her invention commercially. With a discovery of such far reaching implications why has she maintained a low profile ever since? Why has her invention never appeared in commercial development?

On September 1st 1996 an article appeared in the *Sunday Telegraph* entitled 'Breakthrough as Scientists Beat Gravity'. The opening paragraphs of the article read as follows:

Scientists in Finland are about to reveal details of the world's first anti-gravity device. Measuring about 12in across, the device is said to reduce significantly the weight of anything suspended above it.

The claim − which has been rigorously examined by scientists and due to appear in a physics journal next month − could spark a technological revolution. By combating gravity, the most ubiquitous force in the universe, everything from transport to power generation could be transformed.

The Sunday Telegraph has learned that NASA, the American space agency, is taking the claims seriously and is funding research into how the anti-gravity effect could be turned into a means of flight."

That article all but destroyed the career of the Russian scientist Dr. Evgeny Podkletnov, who pioneered the device based on a rapidly spinning superconducting ceramic disc suspended in a magnetic field. His paper never appeared, as planned, in the *Journal of Physics,* and the University of Tandre in Finland stonewalled him denying the research was bona fide. According to Nick Cook, NASA was researching the device before the article appeared in the UK press and continued the research after.

The link between the anti-gravity discoveries of Evgeny Podkletnov, John Searl and Victor Schauberger was spin. John Searle and Victor Schauberger also discovered that spin led to free-energy as well as anti-gravity; which was not apparent from the electrostatic anti-gravity system discovered by T. T. Brown.

The 'over-unity machines', invented by amateur pioneering scientists, have been dismissed on the pretext that it is impossible to get energy out of nowhere. Super-energy research has been opposed, ignored or labeled as pseudo-scientific. The inventors have been discredited crackpots and cranks and their apparatus has been dismissed as perpetual motion machines.

In 1987 Newman had his generator operating as the engine of a car built on a Porsche chassis. Started by a battery the car ran without any input of fuel. However, the American Patent Office refused to grant him a patent for his invention on the grounds that it was, to all intent and purpose, a perpetual motion machine (on the basis that perpetual motion is impossible, alleged inventions of perpetual motion machines are refused patents). Consequently the commercial development of his engine was blocked.

When Trombly attempted to patent his uni-polar generator the U.S. Patent office turned him down on similar grounds. Nonetheless, the U.S. Defense authorities took out a court order against him and threatened him with 10-year imprisonment for infringing secret government research into uni-polar generators.

Schauberger thought he had discovered a means of harnessing atomic energy. Describing his vortex turbines as implosion devices, his research was disrupted by the Allies. Shortly after the end of World War II a group of American military personnel arrived at Schauberger's home in Vienna, seized his apparatus and took him into protective custody. Russian agents were then responsible for destroying his apartment. The American authorities forbade him to resume his research under threat of re-arrest. With the destruction of Victor Schauberger's notes, indicating his construction techniques and the 'special materials' he used along with the confiscation of his equipment, it will never be easy to repeat his experiments.

In 1982, whilst Searl was in the middle of an experiment at his home in Mortimer, Berkshire, England, a group of officials entered the house and confiscated his apparatus. On the basis that he was using electricity without paying bills, he was charged with stealing electricity and fined. The court would not accept that he was generating his own power through a free-energy device. His apparatus was never returned so he refused to pay the fine and was then sent to prison for contempt of court. Whilst in detention a fire broke out at his house, destroying his records. The destruction of John Searl's records has also made it difficult to repeat his experiments.

Another electrical engineer to receive similar treatment, in response to his pioneering work in the field of super-energy, was Wilhelm Reich. Reich had emigrated from Austria to America where he attempted to develop a means of tapping super- energy, which he described as orgone energy. He was able to use his orgone energy devices to dissipate storms and treat diseases, such as cancer. Because of this latter discovery he fell foul of the American Medical Association and Food and Drug Administration. He was committed to prison for infringement of FDA rulings and contempt of court. His books and records were then burnt and the American authorities destroyed his research equipment. Even though, in 20th century America, he was victim to 'book burning' and was proclaimed as a quack and thoroughly discredited, Reich's discoveries were employed in secret research programs and he was encouraged to continue his research into anti-gravity whilst in prison.

In Germany it was the commercial backer rather than the inventor who was arrested. According to a report in *Deutsche Physik*, on May 19th 1992, at 8 a.m. six armed policemen burst into the home of Jurgen Sievers, the Director and General Deputy of a German Company called *Becocraft*. The house was searched and all papers linked to the company were confiscated. On the 15th of June, Sievers was arrested in the street and held in remand at

Koln-Ossendorf. The criminal charge against Sievers and his company was investment fraud in respect to the commercial development of a free energy device, invented by the Austrian inventor, Stephan Marinov.

It has always been difficult for people, involved in the research and development of super-energy generators, to defend themselves in court against a charge of fraud, because science does not allow for the existence of energy outside of the known world. This is one reason why there has never been an open investigation into super-energy as a potential source of power. However, there are reports that suggest there has been secret research into super-energy technology. A free-energy generator was operating, in a secret underground facility, in the United States of America during World War II. According to an intelligence agent who saw it working a decision was taken by higher powers to destroy the apparatus and abandon the research in favour of nuclear energy.

The implications of super-energy, as a source of power, are staggering. Super-energy generators promise a more significant revolution in energy production than the discovery of fossil fuels and nuclear power. It is not difficult to appreciate why government and industry would want to block research in this direction. Fossil fuels and nuclear power are limited, non-renewable sources of energy, which require a large capital investment. This maintains expensive energy production and the generation of vast revenues for those people who finance and control it. Super-energy technology offers an abundance of free energy — after the initial expense of purchasing a generator —and because of their innate simplicity and low production cost, the manufacture of generators could not be controlled. The entire population would have their own free source of power that would negate the need for oil, gas, coal and nuclear power. If bankers, industrialists and governments were to sanction this development they would be cutting their own throats. Not only would they lose their investments in the power utilities, they would see the end of the present transport industries and the road and rail infrastructure that serve them. Super-energy, anti-gravity generators would cause existing forms of transport to become obsolete. With craft that float and run without gas, who would want to drive a car on the overcrowded roads or use a train, or an airplane? Because of this threat to the industrial and economic base of our society, it would be naive to imagine that super-energy vortex turbines would ever be investigated as a source of power in university or government laboratories.

Control of power is one issue, military secrecy is another.

If the prediction of vortex resonance for harnessing super-energy were to be vindicated, the technology would be deployed for military purposes; if this is not already happening. Super powers are obsessed with secret weapon development. Any major technological development in energy production or transportation would be shrouded in secrecy. Disinformation would make it all but impossible to dissect truth from falsehood.

Then there is the possibility that super-energy technology could lead us into other dimensions – more correctly quantum realities – and physical immortality. For the institutions of science and religion that would be heresy.

Conspiracy could be taken as an excuse for the lack of evidence in support of claims for the science of super-energy. Classified is a less loaded word for the same thing;

withholding from we the people, discoveries essential for our future.

If there is a conspiracy the likeliest cause for it would be weapons development. War destroys civilisations, misappropriates resources and maintains poverty. In the midst of world hunger and an ecological and climatic catastrophe the US government awarded Northrop $40 billion in 1981 to develop the Stealth bomber, followed by a comfortable $11 billion a year through the 1990s for the US Air Force to research and develop a range of top secret exotic technologies for the 21st century.

If the science of super-energy were to be used for the benefit rather than the destruction of mankind it would initiate a technological, scientific and theological revolution that would change the world beyond recognition.

Imagine floating cities, miles high, with generators every hundred stories taking out the strain of gravity and producing limitless free energy. Apart from farms and villages, the surface of the Earth could have respite from human civilisation. Mass food production could occur in desert regions thanks to inexpensive desalination made possible by free-energy turbines and anti-gravity technology would allow levitating platforms of growing crops to be moved in and out of optimal climatic conditions. People could fly from city to city in non-polluting, anti-gravity, free-energy craft. There would be no road or rail systems scouring the face of the Earth. Reforestation and reduction in carbon emissions would reduce the greenhouse effect and global warming.

Moving in and out of space time the human population could also make excursions to other planets throughout our own galaxy and beyond. It might be commonplace to welcome visitors from other star systems. Free-energy, anti-gravity craft could bring in resources from other worlds as trade between the planets could be commonplace.

The problem of pollution would be solved as there would be no need for fossil fuels or nuclear energy. Anti-gravity vortex turbines could be used to carry waste into outer space; aimed for the sun for ultimate incineration. Unmolested by mankind much of the Earth could return to its pristine glory as a natural wonderland to be enjoyed for leisure, re-creation and harmonization.

With the promise of physical immortality and reversal of aging made possible by inter-space travel, the Universe is waiting for us to enjoy it. We would have no need to squabble over gods when we are the gods. What reason would we have to fight for a hold on the land when gravity no longer has a hold on us? The struggle for power would be gone when unlimited power is available to all. As a family of humanity we could choose to bury our differences and work together in harmony in this new millennium, to tidy up the Earth, lighten our footstep on her surface and earn a place in the galaxy as responsible and caring citizens.

Alexandersson Olaf, *Living Water: Viktor Schauberger and the Secrets of Natural   Energy*, Gateway Books, (1990).

Cook Nick, *The Hunt for Zero Point*, Arrow 2002

Deutsche Physik, No.7, Morellenfeldgasse 16, A-8010 Graz, Austria

Guston Bill, *Military Power*, Air International, 1/100

Newman Joseph Westley, *The Energy Machine of Joseph Newman*, (1986) - Joseph Newman, Route 1, Box 52, Lucedale, Mississippi 39452, USA.

Roger Silber, *Sacred Geometry Engineering*, Auth., Fairfield, IA 52556, U.S.A

Rho Sigma, *Ether Technology: A Rational Approach to Gravity Control*, Auth.(1977)

Lomas R. *The man who invented the 20th Century – Nicola Tesla.*

Wynniatt C.B., *Energy Unlimited*, Issue 20,1986.

Appendix

"The fact is that physics is not mathematics…stripped of mathematics physics becomes pure enchantment"

Gary Zukav

Appendix to Ch. 2:

Einstein received the Nobel Prize, not for his theory of relativity but for his contribution to quantum theory and our understanding of light. He proved that the electro-magnetic waves of light occurred as particles which came to be known as *photons*. Max Planck, the father of quantum theory, first suggested that energy is radiated in bundles which he called the *quantum*.

The most fundamental unit of energy I represent is a single line of the movement of light. To fit Maxwell's picture of two fields of energy at right angles, I envisage the quantum as a bundle in which there were two lines of the movement of light undulating at right angles.

In my model the quantity of energy is represented by the length of line in the bundle. In a quantum the energy is proportional to frequency. The constant of proportionality came to be named 'Planck's constant' after Max

213

Planck. The greater the number of folds in a bundle of line, the greater the length of the line; this is why the energy in a quantum would be proportional to the frequency of waves it contains.

Planck's constant is represented in physics by the symbol 'h'. If the quantum of energy contains two wave-train lines of the movement of light, a single wave-train line would be delineated by half Planck's constant — ½h

To understand this, imagine ½h as the symbol for the basic energy coinage of the Universe, a bit like the symbol ¢ representing a cent. In normal life we use energy like billionaires whereas the quantum physicist measures energy transactions in cents. Billions of dollars can be treated as integer (whole number) multiples of a cent. If the symbol ½h delineates the fundamental unit of energy in the Universe then obviously in quantum mechanics events could be measured energetically as integer multiples of ½h.

In antimatter creation and annihilation the electron and proton pair are produced from a quantum of gamma radiation therefore each vortex particle would be delineated by ½h. This measure of ½h is called 'quantum spin' but this should not be confused with angular momentum. Momentum laws which apply to vortices – particles of matter – cannot be applied to the energy within the vortex because energy does not have mass. The law of conservation of angular momentum can be applied to the motion of the vortices because the vortices have mass. However, this law cannot be applied to the energy that forms the vortex. For example, the spin of energy forming the vortex is not subject to the angular momentum law but the spin of the vortex causing natural magnetism would be subject to this law.

Magnetism is caused by the movement of charge. Particles with the same charge repel and particles with opposite charges attract. This means that when the energy in interacting vortices is uni-directional there is attraction between them but when it is anti-directional there is repulsion. Mid-way between attraction and repulsion is the state of zero interaction Mid-way between movement in the same and opposite direction is movement at right angles.

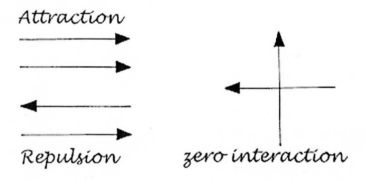

When the lines of movement in sub-atomic vortices of energy cross at right angles they do not interact. This is the basis of electro-magnetism – the distinctness of electric and magnetic fields.

The photon can be described as electro-magnetic radiation insofar as it consists of two fields of radiant energy, each one occurring at right angles to the other. The fact that fields of energy at right angles do not interact is a key to the stability of the photon. Photons radiate through free space for billions of years with no sign of decay. Zero interaction between their constituent fields of energy could be a major contributing factor to their longevity.

The movement of energy causing charge is in and out of the vortex i.e. along its radius.

The spin of electrons in the atom gives rise to natural magnetism. This is the magnetic moment of the electron. The rotational spin of the electron vortex causing its magnetism is effective at right angles to the radial spin causing its charge. The rotational spin causing magnetism is effective at a tangent to the radial spin causing charge.

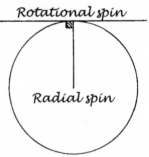

The difference between charge and magnetism can be understood by analogy of whirlpools and tops. The spin of water in a whirlpool represents charge and the spin of a top represents magnetism.

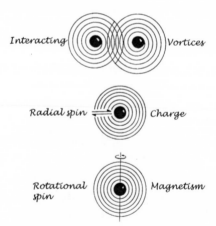

The laws of interaction that apply to radial movements of vortex energy also apply to the tangential movements, e.g. unidirectional interactions of vortex energy cause attraction, anti-directional cause repulsion. The repulsion between two electrons spinning in the same direction would arise from the fact that their rotating vortex energy is anti-directional where it overlaps and interacts.

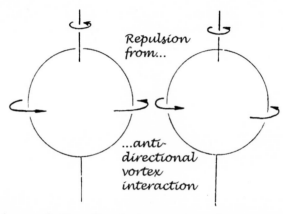

The force of attraction between two electrons spinning in opposite direction would arise from the fact that their rotating vortex energy is uni-directional where it overlaps and interacts.

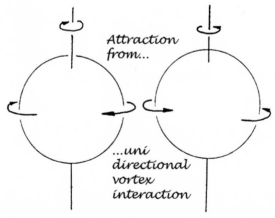

At right angles to the direction of flow of the electrons in an electric current, the motion is effective at a tangent to their concentric spheres of vortex energy. Perpendicular, concentric 'rims of magnetism' surround each moving electron. The concentric rims of magnetism of all the electrons in the current would add up to form the concentric cylinders of magnetism that surround an electric current.

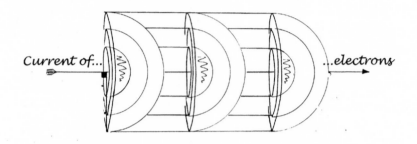

If two current carrying conductors are laid side by side they will attract if the current is flowing in the same direction, but repel if it is flowing in the opposite direction.

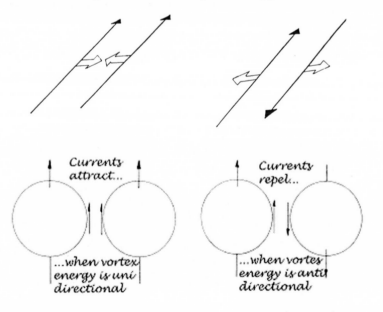

This effect occurs because the concentric rims of vortex energy are anti-directional between the currents flowing in opposite direction and uni-directional between the currents flowing in the same direction.

The intensity of energy, in the vortex, is uniform over the surface of each concentric sphere whereas it varies along the radius. The characteristic of a magnetic field is no change in intensity of vortex energy. The characteristic of an electric field is change in intensity of vortex energy. Michael Faraday discovered that a growing or decaying magnetic field interacts with an electric field. In a growing or decaying magnetic field, the intensity of vortex energy is changing so it would be 'electric' in effect and interact with electric charge. Faraday used this effect is to generate electricity when he invented the first dynamo.

To present my understanding of electricity I draw a parallel between electricity and sound. Sound is longitudinal vibrations in matter. Think of it as compression waves. The compression waves, longitudinal vibrations of electrons in matter produce electricity.

When electrons are free to move in a conductive substance, they can be aligned by the application of a potential gradient across the conductor. Once aligned, movement can be transmitted as vibration from one electron to another down the voltage gradient. Electricity is the passage of 'a shunt' between electrons rather than the passage of the electrons themselves. Whilst the electrons drift slowly in the potential gradient – at approximately three hundred kilometers per hour — the electricity travels at approximately the speed of light — three hundred thousand kilometers per second. The parallel between electricity and sound makes this clear. Sound passes through air at about 1,000 miles per hour whilst, even in a hurricane, the air itself moves at only a fraction of this speed. Sound isn't the flow of the air; it is a vibration passing through the air. Sound is a compression and rarefaction that is tantamount to a shunting of air molecules. Like sound, electricity is the passage of activity through a medium rather than the passage of the medium itself. Sound is the vibration of atomic matter. Electricity is the vibration of sub-atomic matter; it is sub-atomic sound.

Appendix to Ch. 4:

For keys to the knowledge of consciousness go to: http://www.tprf.org/

Appendix to Ch. 5:

In the vortex theory I provide an account for curved space but this is very different from Einstein's. In his general theory of relativity, Einstein based his theory for gravity on the assumption that matter distorts space-time. Einstein predicted that the distortion of space-time around the sun would cause an apparent curvature of starlight as it passes through the 'dip'.

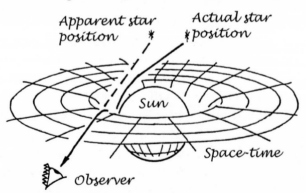

In my theory the curvature of space results from its being an extension of matter rather than something that is distorted by matter.

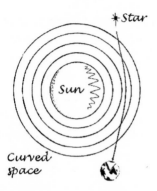

Light would follow the curvature of the concentric spheres of space extending from the sun, much as a car would follow the curvature of the road around a roundabout. This account for the curvature of space shows the same effect on light as Einstein's and unifies space and matter.

Both matter and space are vortex energy. Because the vortex is curved, space is curved. Because the vortex of energy extends beyond our direct perception, bodies of matter can act on each other at a distance. Because the vortex is intrinsically dynamic these interactions can occur without the mediation of any non-material entities.

Einstein assumed that matter distorts space-time. This does not fit with his statement, *"Remove matter from the universe and you also remove space and time."* In that summary of relativity he implied a connection between space and time. How can matter distort space if it is connected to it? Einstein's comment makes sense if space is an extension of matter. Every time a particle of matter is moved a particle of space would be moved with it.

The only arbitrary assumption in the vortex theory is that subatomic particles are vortices of energy. This single axiom explains all the forces and properties of matter — including gravity — in a single uninterrupted process of unfolding logic. The similarity between the equations for charge and gravity support this unification; they suggest that gravity is a vortex interaction rather than an effect caused by the distortion of space-time.

In his special theory of relativity Albert Einstein believed that space 'foreshortens' for bodies moving at velocities approaching the speed of light. For example, we might expect that light emitted from a body, approaching an observer at 100,000 miles per second would appear to have a velocity of $186,000 + 100,000 = 286,000$ miles per second. But Einstein's special theory tells us that the 'mile' for a body moving at 100,000 miles/sec. is shorter than a stationary mile so that the light does not travel as 'far' in a second from the moving body and its velocity again works out at 186,000 miles/sec. This is an arbitrary assumption much like the distortion of space-time by matter. If space is an extension of matter then each observer measures the velocity of light relative to his own space. As this moves with him he would always measure the speed of light to be the same. This configures with Einstein's summary of relativity — that matter and space are connected and so move together – better than his speculation that space foreshortens.

Appendix to Ch. 7:

In 1932, Carl Anderson of the California Institute of Technology made an amazing discovery when he bombarded lead atoms with gamma rays. (Gamma rays are similar to x-rays only they are much more energetic.) In his experiments pairs of particles appeared. One of each pair was an ordinary negatively charged electron but the other was completely alien; it was a positively charged electron which he called a 'positron'.

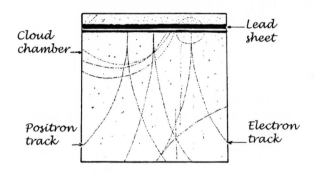

No one had ever seen one of these before though the brilliant Cambridge physicist, Paul Dirac, had predicted their existence in 1928. The positron was the very first particle of antimatter ever to be observed.

From $E=mc^2$ Anderson realised that the energy in the gamma ray had been transformed into matter and antimatter as it passed through the nucleus of a lead atom. The antimatter didn't last long. As soon as the positron met an electron the two particles annihilated and reverted to a pair of gamma ray photons.

Anderson's discovery is easy to explain with the vortex. In his experiment energy in waveform was transformed into vortices. In annihilation the vortices then reverted to waves.

If the gamma ray photon is treated as two lines of the movement of light — undulating as waves of energy — when the two lines of energy flew into the vortex particles in the nucleus of the atom, they were forced into vortex motion. The two brand new vortex-particles then appeared on the other side of the atomic nucleus. One vortex was matter and the other antimatter.

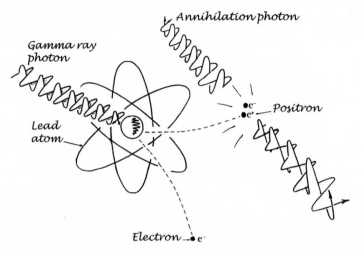

Each fresh subatomic vortex contained an equal half of the quantum of energy brought in by the gamma ray. The new vortex was formed out of a line of the movement of light with the same frequency — transformed from wave into spin. This explained why they both had the same mass and value of 'quantum spin' equal to '$\frac{1}{2}h$'.

The two new particles had opposite charge. This is because the lines of energy in the two vortices were spinning in opposite directions. One direction of spin was into the centre and the other out of the centre of the vortex. Equal in mass but opposite in charge, the particles of matter and antimatter were mirror-symmetrical.

In the annihilation of matter and antimatter the two opposite vortices could be imagined as 'unzipping' or unravelling one another. Because the two opposite whirlpools of light were exactly equal in magnitude they cancelled each other out. Thus the energy bound up in them was released, radiating away as annihilation photons in opposite direction. The energy in the two vortices did not vanish; it changed from particles of matter to particles of light. It was mass and not energy that was annihilated.

The production and annihilation of antimatter confirms photons are half presumptive matter and half presumptive antimatter. In fluorescence and polarization only one of the fields in the photon reacts with matter. In photography only one field of energy in light reacts with the photographic plate. The other field doesn't appear have any part to play in the world of matter. This suggests that James Clerk Maxwell's theory for light as electro-magnetic radiation was an oversimplification. Electric and magnetic fields are part of our world whereas the discoveries of Carl Anderson show that half a photon does not belong in our world at all.

Just prior to their annihilation electrons and positrons, undertake a 'death-dance' called positronium. If the electron and positron are rotating in the same direction, then they attract each other and consequently annihilate sooner. If they rotate in opposite direction they repel and this delays their annihilation. This shows that if the direction of vortex motion setting up charge is reversed, then the magnetic effect is reversed. Particles with the same sign of charge, but spinning in opposite direction, experience a magnetic attraction. When the sign of charge of one of them is reversed then the magnetic effect between them becomes a force of repulsion.

Appendix to Ch. 9:
Cosmic ray research was supplanted by high energy particle accelerators where energies could be controlled. This research was less haphazard than sending photographic plates into high altitude in the hope of harvesting cosmic ray particles. The underlying principle in the accelerators is the same as cosmic research; driving kinetic energy through the vortices in atomic nuclei to produce short lived 'swirls' of mass.

Physicists have been able to increase the energies in

their accelerators to produce a host of new particles heavier than the proton. All the new particles synthesised have one thing in common, they are all very short lived, lasting in many instances, for less than a trillionth of a second. High energy research ratifies the quantum laws of motion.

In high-energy experiments, whilst some of the new particles decay within 10^{-25} sec (the time it takes for light to traverse an atomic nucleus) others lasted far longer. Some, persisting for as long as 10^{-10} sec, were lasting more than a million times longer than expected. Because of their longevity, Murray Gell-Mann — author of quark theory — called them *Strange particles* and *Strangeness* came to be known as a fundamental property of matter.

Strange particles usually decay into more stable particles. Most leave behind them a proton or electron and sometimes even both. Strangeness could be a result of the natural stable vortices acting to delay the operation of the second quantum law of motion. This process could be understood from the analogy of hailstones.

A hailstone forms in a cloud as a result of layers of ice laminating around a seed of dust. The hailstone then falls from the cloud, melts, layer by layer, leaving the particle of dust behind.

According to the vortex theory, as energy is forced through an atomic nucleus in a high-energy reaction, it forms a short-lived vortex around a stable vortex such as a proton, neutron or electron and emerges as a new, unstable particle with a stable particle at its core. On leaving the nucleus, the swirling energy then decays in steps – much as in a hailstone the ice melts, layer by layer. In this process the energy reverts from vortex to wave and radiates away, leaving behind the stable particle — like the hailstone seed of dust. The difference between the proton and the particle of dust in a hailstone is that the proton is itself a vortex of energy – and a very stable one at that. The proton vortex would have a stabilising influence on the temporary vortex that forms round it. A delay in decay of the unstable vortex could occur as a result of the stable vortex at the core exerted a stabilising or holding influence over the unstable vortex of energy swirling around it.

Particles appear in accelerator experiments at critical mass-energy levels in much the same way that electrons occur in the atom at certain energy levels. This principle of discrete energy levels is fundamental to the quantum theory. As physicists increase the energy in their accelerators they generate increasingly more massive particles at successive intervals. The massive particles then decay in steps to reveal new, lighter particles at lower levels of energy where particles tend to appear. Several levels of mass-energy, forming round a proton correspond to what physicists call 'doses of Strangeness'.

The decay of a particle in steps, to reveal lighter particles, is described in physics as 'shedding doses of Strangeness'. If the stable proton, at the core of the unstable strange particle, were to exert a holding influence over it, then as each dose of strangeness is shed, the holding influence over the residual particle would obviously be stronger. This would be because there would be less mass to stabilise and the swirling vortex energy would be closer to the stabilising vortex core. Consequently as each dose of strangeness is shed the new particle should possess a longer life span than the one that went before it. This is evident in the cascade decay of a particle called the Omega-minus.

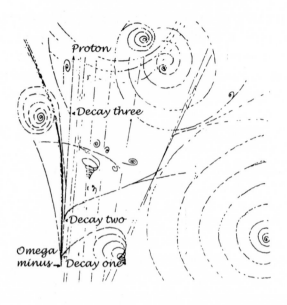

In February 1964, at the Brookhaven National Laboratory in America, physicists witnessed the cascade decay of a new, heavy particle they called the Omega minus. The decay of the Omega-minus substantiates the vortex explanation for strangeness.

In the Brookhaven bubble-chamber photograph of the cascade decay of the Omega minus particle, after each successive decay step the length of track is longer. This indicates that each successive strange particle lasts longer than its predecessors — as predicted in the vortex model. At each stage of the decay, vortex energy is shed as a new, discrete particle, which travels off and decays into radiant energy after a short period of time.

Appendix to Ch. 10:

Compton scattering suggests that the electron is larger than the proton. This effect occurs because the electron has no energy in complete capture. The diameter of a proton is not a measure of the size of the proton. It is the

228

extent of the proton vortex that contains energy in complete capture. This is confirmed by the fact that the range of the strong nuclear force is determined by the diameter of an atomic nucleus in which protons and neutrons are packed tightly together. Like the electron vortex, the proton vortex is infinitely large.

Appendix to Ch. 12:

With over 31.5 million seconds in a year, ($3.15 \times 10^7 + 10^{33} = 3.15 \times 10^{50}$) the difference between the estimated life span of a proton and one of the new particles is a figure in excess of 3×10^{50} (3 with fifty zeros behind it). Mathematicians use the number 10^{50} as a cut off point for probability. If the probability of something happening is less than 1 in 10^{50} then scientists are forced to accept that it never occurs. With a difference in stability in excess of 10^{50}, it can be argued that it is improbable that protons are of similar structure to particles generated in high energy laboratories unless a very good reason is given for the disparity in their life spans. Physicists use a law called the Law of Conservation of Baryon Numbers to account for the disparity in lifespan between a natural proton and the unstable synthetic 'Baryon' particles. (Baryon is from the Greek for heavy.) This law does not tell us 'why' the disparity occurs it simply tells us that is does occur. This is the case with most things in physics. Physicists can usually explain how things work but not why. They have the formulae – as in the case of conservation of baryon numbers — but they lack the fundamental understanding.

In the mid-1960's at a series of experiments were performed at the Stanford Linear Accelerator in California (SLAC). In the SLAC experiments, electrons were accelerated by intense radio pulses, down a three-kilometer long 'vacuum tube' and then targeted on protons in liquid hydrogen. The results of these experiments showed that electrons were being scattered, or bounced back from what appeared to be something small and hard within the protons constituting the nuclei of the hydrogen atoms. From this it was inferred that the protons themselves contained smaller particles. The physicists at SLAC were looking for quarks and naturally concluded that their bombarding particles were bouncing off quarks in the proton. These experiments caused the Quark theory to become accepted into mainstream physics and led to a Nobel Prize for Murray Gell-Mann in 1969.

The SLAC findings appeared to be supported by experiments at CERN, the European Nuclear Laboratory near Geneva. At CERN, protons were being accelerated in vast intersecting rings. At the intersections, energetic protons were directed into head-on collisions. In the high-energy impacts, new particles were formed and came out at right angles to the beams. Again these results seemed to confirm the quark model.

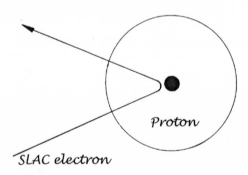

Proton

SLAC electron

However, the SLAC and CERN experiments did not necessarily prove the existence of quarks within the proton; they merely demonstrated that the proton has 'a hard heart'. There could be an altogether different explanation for the 'hard hearted proton'.

As a spherical vortex the proton could be treated as a system of increasingly compact lines of force. Bombarding particles would compress these lines of force, and being vortices themselves would undergo reciprocal compression. The compression of increasingly compact lines of force would result in hard impacts between the colliding particles. In addition to this the increasingly tight 'meson' energy swirling inside a proton would exacerbate the hardness of the proton.

During World War II, physicists succeeded in harnessing the energy locked in the atom to make the most terrible weapons of destruction ever known to mankind. Since then they have been endowed with virtually unlimited budgets for continued nuclear research. Vast sums of money have been spent on building and running massive particle accelerators. Using these, scientists created a host of new particles, which were explained by the quark theory. The theory threw up new difficulties and exciting predictions, which required more research and more powerful accelerators. Each nation, or group of developed nations, wanting to be at the forefront of discovery, was only too happy to spend the equivalent of the gross national product of poorer nations, on a bigger and better accelerator than the 'nation next door'.

Whilst everything in the world is supposed to be made of quarks, not a single quark has ever been observed in a free state — even though no expense or effort has been spared, over the last thirty years, in the effort to find one. Intent 'to throw good billions after bad', governments and scientists have joined together at CERN, in Switzerland, to build the worlds most powerful and expensive accelerator to date. Starting in November 2007 they intend to probe the atom to discover how energy forms mass. The $8Billion they have spent is a waste of money as the Yogis in ancient India already probed the atom with their minds and saw E=mc2 in the vortex. They unlocked the ultimate mystery of creation thousands of years ago for free! We don't need more particle accelerators. Creating a new generation of particles, the new Hadron Accelerator will lead to more elaborations on the theories, which will require more research and eventually the need for an even bigger and more expensive accelerator. To prove the predictions of GUT - the Grand Unified-field Theory - scientists will need an accelerator as big as the solar system.

High-energy research has created an endless cycle in which research physicists manufacture problems for theoretical physicists to solve and they, in turn, predict yet more problems to research. Whilst this elaborate game ensures employment for scientists and their army of technicians, it produces information which is of little relevance or value to the tax payers who have to foot the enormous bill: ...*In a sad world of homeless and starving people there once lived a mad professor who believed in flying pigs. No one had ever seen a flying pig but the mad professor managed to convince all the Universities in his world that hunting for flying pigs was the most valuable line of research they could undertake. As the greatest minds in the nutty world caught flying pig fever, the developed nations fell over themselves in the frenzy to build bigger and ever more elaborate flying pig blasters. No expense was spared. Money taken from the people, that could have been used to improve the quality of their lives, was squandered in the search, but after decades of hunting no one spotted a single flying pig. After all that effort you would have thought that the learned professors would have accepted that their flying pig theory was cranky, but not willing to admit to being cranks, they just went on building ever bigger fly-*

ing pig traps, guns and telescopes.

Another nut case, in that crazy world, was the mad professor of puddings! A student in his department of nutrition baked plums in a pudding then ran round shouting, "Eureka, I have just invented plum pudding!"

The professor was not impressed. He snorted over his spectacles, "That is not a plum pudding. That is a Blackforest gateau."

"But," argued the downcast student, "I mixed plums into my pudding before I baked it and when I weighed it, the weight was that of the plums and the pud and at the end, when I shook the pudding, out came a plum. It has to be a plum pudding!"

The professor became really angry. "You stupid student, have you learnt nothing of what I have taught you? When you bake plums in a pudding the ingredients completely change their identity. Due to the interaction of a weak cooking force, the plums change into cherries and pudding becomes a Blackforest gateau."

"Well how do you explain the plum that fell out of the pudding?" retorted the student; "There are no plums in a Blackforest gateau!"

"The gateau is unstable," replied the professor, "after a few minutes it falls apart into its original ingredients - plums and pudding."

How could the student argue? He was speaking to none other than the President of the Royal Society of Puddings. If he wanted to make it in his world and get a good career in the cake and pudding industry, he had to accept that a pudding full of plums was a Blackforest gateau...

Electrons are very light and bear a negative charge. They could be likened to the plums in the story. Protons bear a positive charge and are 1836 times as massive as electrons. They could be likened to the pudding mass. The neutron is neutral in charge and can be formed out of an electron and proton in a process called K-capture. The neutron has slightly more than the sum mass of an electron and proton and in a process called Beta decay it falls apart into an electron and a proton. From this you would imagine the neutron must be a bound state of electron and proton, represented by the plum pudding. However, I don't recommend you suggest this to a professor of physics. He would think you are quite stupid to jump to that obvious conclusion. *"Neutrons are not electrons bound to protons,"* he would insist. *"The electron and proton lose their identity as they come together to form a neutron but regain their identity again when the unstable neutron falls apart."*

If you stayed awhile to listen he might tell you that the proton is made up of two up-quarks and one down-quark. He would then explain that the neutron is made up of two down-quarks and one up-quark and go on to say that when a proton interacts with an electron, a force called the weak nuclear force comes into play. This transforms an up-quark into a down-quark and the electron into an elusive particle called an anti-neutrino.

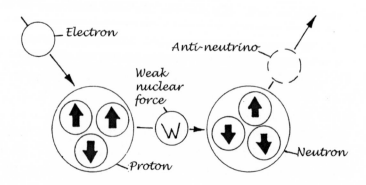

If you asked him to elaborate, he might tell you that in this process the electron and proton would have ceased to exist, their place being taken by the neutron and anti-neutrino. If you were alarmed at this he would reassure you that they come back into existence if, with the release of a neutrino, the process is reversed. However, I suggest you look for an exit sign, if he begins to talk about fermions, quantum-spin, Z and W particles, and make a run for it should he begin to support his arguments with Fermi-Dirac statistics. Should you get completely lost and find yourself wondering what quark is, you should find it down at your local supermarket somewhere between the yogurt and the cream cheese!

Meanwhile... back in the nutrition department of the crazy world, all was not going well with the mad professor of puddings. Another student baked plums in a pudding then licked it. He tasted a plum and so established that the plums did not lose their identity when baked in a pudding. The mad professor was very embarrassed by the arrival of that awkward fact. He ranted and raved and made it a rule that food was never again to be tasted in the lab. And then he used his great prestige and authority to ensure that when the result of the lick-experiment was printed in the textbooks, it was set in small type so that hopefully, it would be overlooked as a totally unimportant fact....

The 'lick-experiment' highlights the discovery of a number of awkward facts about the neutron. In 1957, Smith, Purcell and Ramsey discovered that a neutron has a slight electric dipole moment. This means the neutron is not entirely neutral. On one spot it displays a minute negative charge, in the order of a billion, trillion times weaker than that of a single electron. This suggests that the neutron is a bound state of opposite charges, which mostly cancel each other out (This occurs in an atom, which is electrically neutral because it contains equal numbers of oppositely charged particles.)

When first I was shown the information about the electric dipole of a neutron, it was in a short paragraph and set in small type. That aroused my suspicions. I could smell a rat. There had to be some reason why physicists were so adamant that a neutron was not an electron bound in a proton, in the face of obvious evidence to suggest that it was and the dipole moment appeared to be an awkward fact. When I came across it again, in the physics library of Exeter University, it was in a textbook on nuclei and particles by Emilio Segre. Segre gave the figure for the electric dipole moment of the neutron as being equal to the charge on an electron x 10^{-20} (-0.1 +/- 2.4) and followed it with the comment, *"...hence this moment could be exactly zero, in agreement with the theory."*

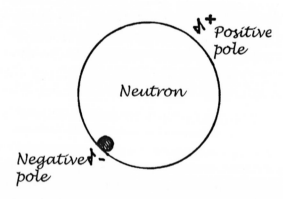

Whilst the large margin of error does allow the 1957 measure to be taken as 'inconclusive' the way this comment was phrased suggested to me that Segre wanted the measure to be dismissed as an unimportant fact in line with the normal practice in mainstream science to dismiss as 'inconclusive' research that challenges accepted theories or the materialistic worldview.

The presence of charged particles in the neutron is supported by the magnetic moment of a neutron i.e. 1.91 nuclear magnetrons. The neutron, if it were a truly neutral particle, would have no magnetic moment because the magnetism of a particle is created by the spin of its charge. A particle cannot have a magnetic moment unless it has a charge. The magnetism of a neutron adds support to the view that it is a bound state of two opposite charges, which mostly cancel each other out rather than a single particle with no charge at all.

Another piece of evidence to support the view that electrons are bound within the neutron came from a discovery made in 1956 by the American physicist Chien Shiung Wu. Wu lined up the nuclei of radioactive atoms in a magnetic field so they were all spinning in the same direction and observed that more electrons were emitted in one direction than in another. This experiment showed that in beta decay electrons are emitted directionally as if they emerged from a specific site on a neutron.

To quote Harald Fritzsch, "We *do not understand why the neutron is heavier than the proton. Indeed an unbiased physicist would have to assume the opposite by the following logic. It is reasonable to think that the difference in mass between the proton and neutron is related to electro-magnetic interaction since the proton has an electric field and the neutron does not. If we rob the proton of its charge, we would expect the neutron and the proton to have the same mass. The proton is therefore logically expected to be heavier than the neutron by an amount corresponding to the energy needed to create the electric field around it.*"

There is overwhelming evidence to support the view that an electron is bound to a proton to form a neutron. Against this, physicists argue that a neutron has the same value of quantum spin as a proton or an electron. They say that if it is a bound state of an electron and proton it should display the quantum spin of them both. It is obvious that if a light electron were bound by a massive proton, it would be all but 'lost from view' so that its quantum spin would be hidden. The inertia of the proton, conferred by its greater mass, would account for the neutron appearing with the spin of the proton. Physicists may contend that the law of Conservation of Angular Momentum does not allow for the spin of an electron to be lost but they allow for conservation laws to be upheld in the formation and decay of unstable particles — so long as nothing is gained or lost in the overall process. Spin would be conserved in the overall process of formation and decay of the unstable neutron because an electron going in would have the same spin as an electron coming out.

Physicists are clinging to straws in their arguments against the obvious structure of the neutron. It is clear they are attempting to cover something up. The reason they are so determined that a neutron is not a bound state of electron and proton is that the bound state theory for the neutron threatens the Heisenberg uncertainty principle, the foundation of quantum mechanics; the sacred cow of modern physics!

The neutron, discovered five years after the Uncertainty Principle was proposed, offered an ideal experimental test for the principle. If a neutron is treated as an electron bound to a proton, the position and momentum of the electron could be defined with a high degree of certainty. The position of the electron would be somewhere within a space defined by the estimated size of the neutron. Its momentum could be ascertained from the energy locked up in the mass of the neutron that exceeds the sum mass of an electron and proton e.g. the rest mass of an electron is 0.911×10^{-30} kg. The rest mass of a proton is 1672.62×10^{-30} kg. The rest mass of a neutron is 1674.92×10^{-30} kg; therefore, it has a mass of 1.389×10^{-30} kg in excess of the sum mass of an electron and proton. This is equal to 1.5x the rest mass of an electron. In the neutron, an electron could not possess a momentum in excess of that allowed by the energy locked up in 1.389×10^{-30} kg of mass.

If Heisenberg's principle is applied to an electron confined in the space of a neutron, the high degree of certainty in its position demands of it an enormous indeterminacy in momentum. The Uncertainty Principle predicts that electrons bound in a neutron could have momentum values requiring velocities as high as 99.97% of the velocity of light. To reach that order of velocity, the electron would have to possess energy up to the equivalent of approximately 40x its rest mass. If the Heisenberg principle were valid then neutrons — as bound states of electrons and protons — would occur with a wide range of masses reflecting the indeterminacy of the momentum of their constituent electrons e.g. to range from 1673.5×10^{-30} kg to 1710×10^{-30} kg. This would reflect a range of electrons with zero momentum to electrons with velocities approaching the velocity of light. However, all neutrons possess the very precise mass of 1674.92×10^{-30} kg. This indicates that there is no detectable, indeterminacy in the momentum of their constituent electrons.

The theories of quantum mechanics cannot stand if the neutron invalidates the Heisenberg Uncertainty Principle. If the neutron is a bound state of electron and proton then the outstanding experimental success of quantum mechanics will have done little more than reveal just how effective science can be in establishing myths. It is hardly surprising that when the neutron appeared, as a black swan, over the quaint little pond of quantum mechanics, instead of welcoming its arrival — as scientific integrity would demand — physicists tried to shoot it. Had they allowed it to land, Heisenberg's principle would have been a dead duck and the virtual force-carrying particles, their darling little ducklings, would have certainly drowned. Failing to submerge it, they muddied the waters with incomprehensible math, whitewashed it with quark and with a flap about quantum spin, chased it away.

In the pre-scientific era, theories were built up by heaping speculation upon speculation. Medieval theologians were speculating that if an angel were chopped in half, two angels would emerge because new halves would grow on the split ends. Philosophers were adding epicycles to Ptolemy's complex description of the solar system to accommodate new discoveries in astronomy whilst alchemists were involved in an endless desperate search for the philosopher's stone. History would appear to be repeating itself in physics as speculation is heaped upon speculation and the endless search for hypothetical particles becomes ever more desperate. Meanwhile, protons, neutrons and electrons, the real stuff of matter, remain unexplained. It is the story of the Emperor's New Clothing all over again. Quarks and the virtual particles of quantum mechanics are phantoms that don't exist, but few people in physics care to admit to this because there are too many jobs, academic reputations and Nobel Prizes at stake. However, there is no future for science if awkward facts are swept under the carpet to save the theories.

In his 1980 Inaugural Lecture, *'Is the end in sight for Theoretical Physics'*, Stephen Hawking said that because of the Heisenberg Uncertainty Principle, the electron could not be at rest in the nucleus of an atom. The fact is that electrons are bound within atomic nuclei and it is precisely because of this undeniable fact that the end is in sight for theoretical physics.

The awkward facts about the proton and neutron could mean the end of theoretical physics, as we know it. However, this may not be a bad thing. Appearing as pillars of truth and harbingers of change, the evidence of these natural particles will enable us to expunge the myths of modern physics and clear the way for a new approach to quantum theory and a completely new understanding of the Universe.

Appendix to Ch. 18:

Much like a team of Inquisitors, the editor of Nature, John Maddox descended on Dr Jacques Benveniste's lab with the 'quack-buster' journalist Walter Stewart and professional magician James Randi. They were determined to reveal Benveniste as a fraud. In the magazine *Nature* the experiments were denounced as delusion. Sceptics and scientists were quick to join the fray with cries of 'dubious science', a 'cruel hoax' and 'pseudo-science'. These accusations were more appropriate to Maddox and his team than Benveniste, nonetheless Benveniste's reputation was ruined. He lost his position as director at the French National Institute for Health and Medical Research and found it difficult to find a position in scientific institutions elsewhere. Undismayed by his public denouncement as a quack and a fraud, Jacques Benveniste continued his research that established a sound basis in science for homeopathy.

Appendix to Ch. 20:

There is speculation that flying saucers were developed by Germany during World War II and subsequently became part of an Allied top-secret military development in the post war years. This would have accounted for the spate of UFO sightings in Europe and America and elsewhere in the world immediately after the war. However, this theory does not account for the numerous UFO sightings during the war by Allied and German pilots and crew. There is no doubt that the Germans were ahead of the Allies in developing anti-gravity flying discs and jet aircraft. However, the evidence suggests they were still in prototype state when the war was over. If Germany was responsible for the huge numbers of *'foo fighters'* sighted they would have used the technology to gain air superiority and win the war.

Calder Nigel, *Key to the Universe: A Report on the New Physics* (1977), BBC Publications

Cook Nick, *The Hunt for Zero Point:* (2001) Arrow Books.

Fritzsh Harald, *Quarks: The Stuff of Matter* (1983), Allen Lane.

Gamow George, *Thirty Years that Shook Physics,* Heinemann.

Hawking Stephen, *A Brief History of Time* (1988) Bantam Press.

Hawking Stephen, *Black Holes and Baby Universes* (1993) Bantam Books.

Leggett A.J., *The Problems of Physics,* (1987) Oxford University Press.

Segre Emilio, *Nuclei & Particles* (1964) Benjamin Inc.

Zukav Gary, *The Dancing Wu Li Masters* (1979) Rider.

Index